Human Biology
Laboratory Manual

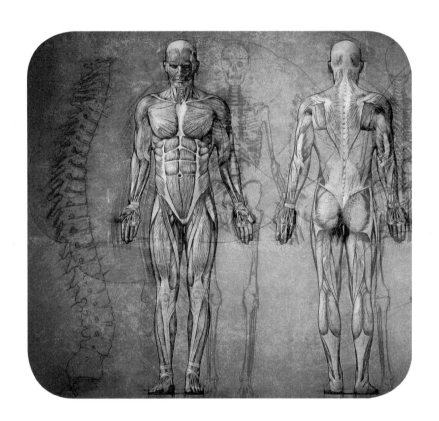

Dr. Crystal Anthony, PT

www.kendallhunt.com
Send all inquiries to:
4050 Westmark Drive
Dubuque, IA 52004-1840

Copyright © 2018 by Kendall Hunt Publishing Company

ISBN: 978-1-5249-5232-7

All rights reserved. No part of this publication may be reproduced, stored in a retrieval system, or transmitted, in any form or by any means, electronic, mechanical, photocopying, recording, or otherwise, without the prior written permission of the copyright owner.

Published in the United States of America

CONTENTS

Lab 1	Laboratory Safety	1
Lab 2	Scientific Method	5
Lab 3	Metric System	17
Lab 4	Chemistry	27
Lab 5	Microscopes and Cells	47
Lab 6	Diffusion and Osmosis	61
Lab 7	Tissue Organization: Histology	75
Lab 8	The Heart	89
Lab 9	Blood Cells	99
Lab 10	Blood Typing	105
Lab 11	ELISA	109
Lab 12	Organs, Cavities, and Digestion	121
Lab 13	Respiratory System	127
Lab 14	Respiratory Response	131
Lab 15	Urinalysis	137
Lab 16	Skeletal System – Bone Lab	145
Lab 17	Muscular System: Skeletal Muscles Lab	149
Lab 18	Brain Models and Dissection	153
Lab 19	Introduction to EMG	157
Lab 20	Human Senses	163
Lab 21	Homeostasis	175

Lab 22	Reproduction and Embryology	183
Lab 23	Cell Division	191
Lab 24	"Genes in a Bottle" Cheek Cell DNA Extraction – Capture Your Genetic Essence in a Bottle	199
Lab 25	Mendelian Inheritance	207
	Practical Exam Sheets	213

Laboratory 1

Laboratory Safety

Objectives

- Learn how to be safe in the laboratory
- Discover where the lab safety items are in the lab room
- Be prepared in case of accidents and spills

To learn about lab safety complete the following:

- Read the following procedures carefully
- Complete the scavenger hunt items (marked by open circles) and label the classroom map on the last page with your discoveries.
- Complete the short safety quiz at the end to test your safety knowledge.
- Sign at the bottom and return the completed assignment to your instructor

Laboratory Safety Procedures

Your safety in the laboratory is very important to us. Preliminary lab instruction starts promptly at the beginning of the lab period. It is the student's responsibility to arrive on time, listen to the instructions, and follow the directions and guidelines given. The student must also use responsible classroom behavior: immature behavior, horsing around, or other childish acts are a major cause of accidents and have no place in a laboratory. The following are rules and guidelines for a safe and injury-free lab experience and are by no means all-inclusive. The student is responsible for using common sense at all times while in the laboratory.

- Notify your instructor if you have any medical condition (pregnancy, allergies, disabilities, immunological disorders, etc.) that may require special precautionary measures in the laboratory.
- DO NOT eat, drink, smoke, apply cosmetics, etc., in the lab. Leave food and drink outside of the lab or keep it put away inside your backpack.
- Items (coats, books, etc.) that are not required for the lab should be stored in an out-of-the-way place (under the lab table or in cubby). Desk space is minimal and must be reserved for essential equipment and your lab manual.
    - ○ Find and indicate on the classroom map where the cubby holders are.

- Wear closed toe shoes in the laboratory at all times. We will be working with glass equipment that could break and if your toes are exposed you will be running the risk of dangerous cuts. If you do break glass let your instructor know and carefully dispose of the broken glass in the glass waste container.
    - Find and indicate where the glass waste containers are on the classroom map.
- A lab coat or apron, goggles, and gloves must be worn for certain lab activities. It will shield your clothing from contamination and shield expensive clothing from stains and other reagents used in the lab. Lab coats, goggles, and gloves will be provided when needed.
- Wash your hands and arms thoroughly with soap and water EVERYDAY before leaving the laboratory to be sure that you are not taking chemicals or organisms with you that could later contaminate your food.
    - Indicate on the classroom map where the sinks with soap and paper towels are.
- When using open flames or chemicals, long hair MUST be secured in a ponytail. Turn off all flames when not in use – NEVER walk away from a lit burner. Notify instructor or assistant immediately if you smell gas.
- Completely plug in electrical devices to avoid electrical shock.
    - Indicate on the classroom map where the CLOSEST gas tap (the gas is shut off so don't even try it) and electrical outlet to you is.
- Be aware of hot items in the lab; always assume that hot plates are HOT, and boiling water baths are HOT. Wear appropriate gloves and NEVER use your fingers to remove anything from a hot plate or a boiling water bath.
    - Indicate on the classroom map where the location of the water bath.
- Immediately report any accident, injury or spill to the instructor or assistant. Cuts or abrasions suffered in lab need to be IMMEDIATELY treated with antiseptic. Cover any cuts you may have before coming to lab. ASK if you need a Band-Aid.
    - Find and indicate on the classroom map where the first aid kit is located.
    - Label where the telephone and emergency procedures are located.
- If hazardous chemicals enter your eye, have someone notify the instructor and (with assistance) flush your eyes out in the eye wash station for 5 minutes. If hazardous chemicals get on your clothes, remove them immediately, have someone notify the instructor, and use the shower station to wash off the chemicals for 5 minutes.
    - Find and Indicate on the classroom map where the eye wash and shower stations are located.
- In case of fire notify your instructor immediately and be aware of the location of the fire extinguisher and fire blanket. The fire extinguisher is used to put out open flame whereas the fire blanket is wrapped around a burning individual.
    - Indicate on the classroom map the location of the fire extinguisher and fire blanket are located.

- In this laboratory we will be culturing bacteria and fungi in experiments. Treat ALL microbial cultures as if they were human pathogens. Place used cultures in red biohazard bags. If your skin is exposed to the microbes, wash the affected area with warm water and soap immediately. Wear gloves. Use gloves and clean the countertops with provided cleaner before and after class starts, act as if there are pathogens on the table that need to be cleaned. We call this using Universal Precautions to prevent spread of disease.
- DO NOT remove any specimens, cultures, chemicals, reagents, or any other items from the laboratory at any time. This is for safety and for protecting our supplies. Taking things that don't belong to you is stealing.
- NEVER wash material down the sink drain unless told to by your instructor or lab assistant.
- At the end of the lab period, rinse out, wash and dry used glassware, dispose of used items in the properly labeled receptacles (slides, sharps, dishes, etc.), and return all lab materials and equipment to the lab cart. **If you steal equipment from the lab you will fail the class and the police department will be notified.**
    - Indicate on the classroom map where the garbage cans are.

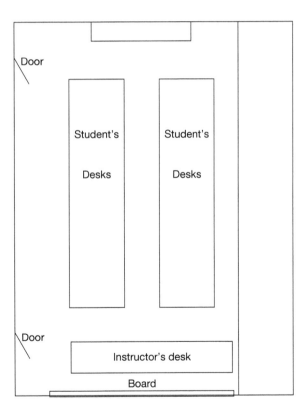

Answer Sheet

Lab 1: Laboratory Safety Questions

Name: _____ **Section:** _____

1. What should you wear for a laboratory experiment involving hazardous chemicals?

2. If ANY injuries or spills occur, who should be notified immediately?

3. How is broken glass disposed of properly?

4. Where should backpacks and coats be stored during lab?

5. True or False? It's fine to eat and drink in the lab.

6. True or False? Horseplay is acceptable in the lab.

7. True or False? It is important that you read you lab prior to coming to class.

8. True or False? All chemicals can be poured down the sink when you are done with your experiment.

9. True or False? You don't have to worry about washing your hands after every lab, only the ones with bacteria.

10. True or False? The instructional aide will pick up after you so you don't have to put anything back on the lab cart

_____ _____
Print Name   Section #

_____ _____
Student Signature   Date

This page must be kept on file in the Biology Department

Laboratory 2
Scientific Method

Objectives

- Describe and perform the steps of the scientific method
- Explain the process and purpose of an experiment
- Design and perform an experiment

Introduction

To learn how scientific ideas develop, it is necessary to understand the process of the scientific method. In order to understand the world around us, and ourselves, we must learn to observe, question, investigate, and think critically.

We begin with an observation that leads us to a question. We create a hypothesis, which is a possible answer to our question. Once data are collected and analyzed, conclusions are formed. The hypothesis is supported or rejected. The cycle continues with changes and refinements. More observations can be made to further support and strengthen the results.

Adjustments to the initial question and/or modification of the hypothesis may be made. Changes in the experimental conditions provide us with more information and additional conclusions. The cycle continues as researchers continually build on previous research and ideas. The diagram gives a general idea of the process.

The Scientific Method as an Ongoing Process

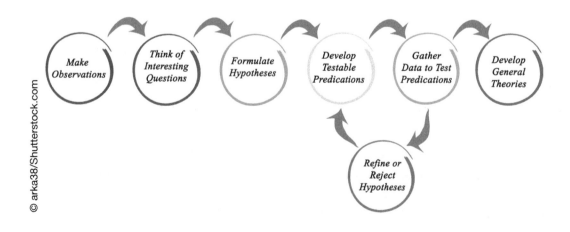

I. Scientific Method

Read through the steps of the scientific method, then complete chart and answer questions.

1. **Observation:** The scientific method begins with an *observation*. A scientist observes things using their own senses and knowledge base. The observation of some event gets the mind thinking about how things work or why certain things occur. You will be observing the human body.

2. **Question:** We then *ask questions* about these observations. The more carefully we observe, analyze, and research what other knowledge has been collected by other scientists, the more specific and applicable our question will be.

3. **Hypothesis:** Next, a *hypothesis* is developed. We use *inductive reasoning* to find a possible answer to the question. Inductive reasoning takes into account many observations, recognizes patterns, makes generalizations, and formulates a reason (i.e., possible answer to the question). For example, I observe my children start running around the house 5 minutes after eating candy. I observe a pattern because 30 times after they ate candy they ran around the house. I can predict that they will run around the house the next time they eat candy. My hypothesis is that the candy is causing the children to run around in circles. A good hypothesis must have two characteristics: (1) must be able to account for all the available data/information, and (2) must be testable. A testable hypothesis is one that can be subject to experimentation.

4. **Prediction:** A *prediction* is created allowing the scientist to design an experiment that tests the hypothesis. We use **deductive reasoning; "if-then"**. I formulate a testable hypothesis: If candy peaks energy levels causing children to run around, then I will see a blood sugar spike 5 minutes after the candy is eaten.

5. **Perform experiment; collect and analyze data:** We *perform an experiment* to test the prediction. A simple controlled experiment creates two groups: a **test (experimental) group** gets the treatment and a **control group** does not. More complex studies involve more than one experimental group. In any test, there are three kinds of variables. The **independent variable** is the treatment or condition under study. The **dependent variable** is the event or condition that is measured or observed when the results are gathered. The **controlled variables** are all other factors, which the investigator attempts to keep the same for all groups under study. The results of the experiment are referred to as data and are tabulated in a graph and/or table. Mathematical equations are used (a branch of mathematics called *statistical analysis*) to analyze the data.

6. **Make a conclusion:** To *make a conclusion* you use the results of the experiment to determine whether or not you can **support the hypothesis**. If the results do not support your hypothesis, then you **refute the hypothesis**. You then go back to the drawing table, modify your experiment and/or modify your hypothesis, then experiment again.

There is absolutely no "proving" anything in science. We can never prove the hypothesis simply because there may be other factors we are not considering. Even in the best experiments, a scientist cannot consider all possible variables. Because we cannot claim to know all there is to know in the world, there may be another explanation for what we have observed. The more research and experimentation that is conducted, the more confident we are that our original hypothesis is on-track. This is where that idea of replication comes in. Repeating an experiment over and over can either give validity to a hypothesis or it can refute it.

Hypotheses that are widely tested and supported over and over again indicate a fundamental pattern in nature called a theory or law. A **theory** in common culture is very different from a theory in science. A **scientific theory** (like the theory of evolution by natural selection) started as a hypothesis but has been *tested and supported over and over again*. It is a widely accepted explanation for an observation. **Laws** are theories that continue to *be supported by evidence gained from experimentation* (e.g., the *law* of gravity). However, like hypotheses, theories and laws can be modified or even discarded in the light of new knowledge.

Finally, the experiment must be documented. Typically, a researcher will have their work published in a scientific journal so that others can read their work. The published document will include all pertinent information to allow another researcher to conduct the same (or very similar) research to see if the same results are achieved. Published scientific research contributes to all of the scientific body of knowledge and drives further studies to continue to expand on that knowledge.

"I wish you had chosen a more pertinent educational issue than 'Do Dogs Actually Eat Homework?'"

LINE GRAPHS

Introduction

Line graphs like the one below can be used to show how something changes over a time period or over a concentration range. Line graphs are great for comparing different conditions or experimental groups and the effect on a specific variable over time. Line graphs are good for plotting data that have peaks (ups) and valleys (downs), or that was collected in a short time period. The following pages describe the different parts of a line graph. Refer to the example line graph below to see what the different parts correspond to on an actual graph.

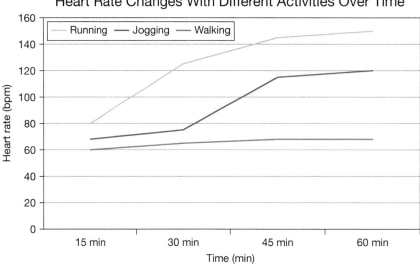

The Title

The title offers a short explanation of what is in your graph. This helps the reader identify what they are about to look at. It can be creative or simple as long as it tells what is in the graph. The title of the graph above tells the reader that the graph contains information about the heart rate changes when engaging in different physical activities during a 60-minute period of time.

The Legend

The legend tells what each line represents. Just like on a map, the legend helps the reader understand what they are looking at. This legend tells us that the green line represents running, the red line represents jogging and the blue line represents walking.

Y-Axis

In line graphs, the y-axis runs vertically (up and down). Typically, the y-axis has numbers for the amount of stuff being measured by your experiment. The y-axis usually starts counting at 0 and can be divided into as many equal parts as you want to. In this line graph, the y-axis is measuring heart rate. This is the dependent variable, what you are measuring. (Does the type of exercise have an impact on heart rate?) (Does time have an impact on heart rate?) Make sure to indicate the units of measurement in your axis label (i.e., dollars, centimeters or cm, grams or g, etc.)

X-Axis

In line graphs, like the one above, the x-axis runs horizontally (flat). Typically, the x-axis has numbers representing different time periods or other continuous variables that are not affected by the experiment. In this line graph, the x-axis measures time. Make sure to indicate the units of measurement in your axis label (i.e., seconds or sec, minutes or min, etc.)

The Data

The most important part of your graph is the information, or data, it contains. Each data point is plotted on the graph based on its value for the x-axis variable and y-axis variable. After plotting the data points they are connected with a line. Line graphs can present more than one group of data at a time. In this graph, three sets of data are presented (for walking, jogging, and running).

Bar Graphs

Sometimes line graphs are not needed and a simple bar graph is more appropriate. *Note*: Just because bar graphs are easier, does not necessarily make it the best choice. Take time to determine which graph is best to use, you will get marked off if you use a bar graph when a line graph is the better choice. This is a good reference for determining the type of graph to use: http://www.biologyforlife.com/graphing.html

Bar graphs are best for things that are one time, not continuous over time. Used when tallying things such as the number of people with a given feature.

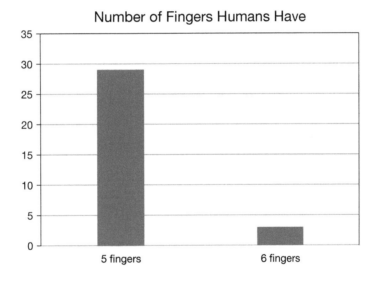

Lab 2: Pre-Lab Questions

Name: _____

Read through **all** of the lab handout and answers these questions **before** coming to class.

1. The variables that investigators try to keep the same for both the experimental and the control groups are
   a. Independent
   b. Controlled
   c. Dependent
   d. A and C

2. The first step of the scientific method is to
   a. Ask a question
   b. Construct a hypothesis
   c. Observe carefully
   d. Formulate a prediction

3. In an experiment the subject or items being investigated are divided into the experimental group and
   a. The non-experimental group
   b. The control group
   c. The statistics group
   d. The blue man group

4. Variables that are *always* different between the experimental and control groups are
   a. Independent
   b. Controlled
   c. Dependent
   d. Reliant

5. The results of an experiment
   a. Don't have to be repeatable
   b. Should be repeatable by the investigator
   c. Should be repeatable by other investigators
   d. Must be both B and C

6. The detailed report of an experiment is usually published in a
   a. Newspaper
   b. Book
   c. Scientific Journal
   d. Magazine

7. What is the importance of statistical analyses?
   a. They can reveal whether or not the data have been fabricated
   b. They can only be used to support the hypothesis
   c. They can be used to determine whether any observed differences between two groups are real or a result of chance.
   d. All of the above

Use this information for Questions 8 and 9:

You conduct a clinical trial to test whether a new drug relieves the symptoms of arthritis better than a placebo. You have four groups of participants, all of whom have mildly painful arthritis (rated 7 on a scale of 1 to 10). Each group receives a daily pill as follows: control (group 1)– placebo; group 2–15 mg; group 3–25 mg; group 4–50 mg. At the end of 2 weeks, participants in each group are asked to rate their pain on a scale of 1 to 10.

8. What is the independent variable in this experiment?
    a. The amount of pain experienced at the start of the experiment
    b. The different drug treatment groups
    c. The degree to which pain symptoms changed between the start and end of experiment
    d. The drug itself

9. What is the dependent variable in this experiment?
    a. The amount of pain experienced at the start of the experiment
    b. The different drug treatment groups
    c. The degree to which pain symptoms changed between the start and end of experiment
    d. The drug itself

Answer Sheet

Lab 2: Scientific Method

Name: _____ Section: _____

1. Fill in the chart below. Ask a question based on each of the observations listed below. Create a hypothesis based on your question. Devise a prediction that is testable for experimentation. **In the last row you need to state one observation you have about the human body and then fill in the remainder boxes.**

TABLE 2.1 Human Body Observations, Questions, Hypotheses, Predictions

Observation	Question	Hypothesis	Prediction
People can hold their breath for a short time	How long can people hold their breath for after doing exercise?	People can hold their breath for 20 seconds after doing exercise.	If a person does 50 jumping jacks, then they can hold their breath for 20 seconds.
People have two eyes			
People have two ears			
Your observation:			

2. Pick one of the hypothesis to test. Write **the predictions you will be testing on the line below.** Then, **design and conduct the experiment** with your classmates.

   _____

3. Define the terms below and then describe the variables involved with testing the prediction you chose from Table 2.1

- Independent variable:

  _____

  _____

  _____

  _____

- Dependent variable:

  _____

  _____

  _____

  _____

- Controlled variables:

  _____

  _____

  _____

  _____

4. Show your data in the chart below. Not all boxes need to be used. **Clearly label and note the units of measurements**, if applicable. Create a larger chart and attach the extra page if more room is needed.

Data:

5. Use the data above to construct your graph below. Decide which variable will go the x-axis and which will go on the y-axis. After plotting the data points connect them with a line or color in the bar graph.

Be sure to include all of the parts of a line graph explained in introduction: Title (descriptive but concise), x-axis labeled with units, y-axis labeled with units and the Legend.

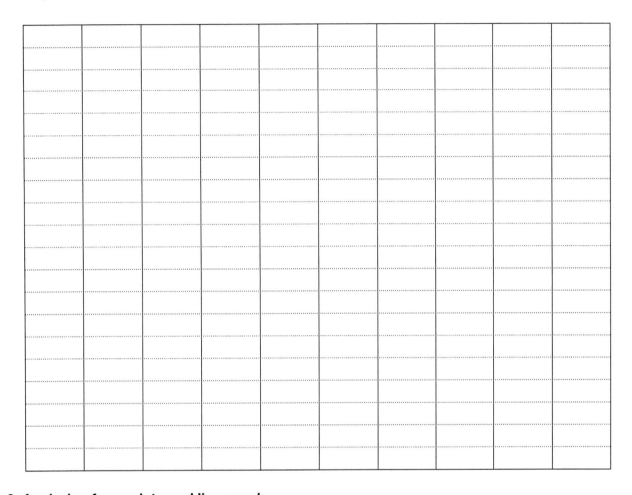

6. Analysis of your data and line graph

Write a conclusion for your experiment, based on the data you gathered and the results of your experiment. Make sure to reference your data and relate it to your hypothesis.

Conclusion:

___
___
___
___
___
___
___
___
___
___
___

Laboratory 3

Metric System

Objectives

- Become familiar with the metric system
- Make observations
- Construct hypotheses
- Make measurements
- Test hypotheses
- Evaluate data

Background: Measurements in Science

The metric system is perhaps the most widely adopted system of measurement in the world and is, therefore, the system used by all scientists. Even in the United States, where there has been some resistance to utilizing this system, there has been an increase in the use of metric units. We are always encountering metric system units: soda is packaged in containers that are measured in milliliters or liters (instead of cups, ounces, or gallons), pearls are measured and sold by their diameter in millimeters (not inches or yards), the medicine you take to alleviate a toothache or to combat another ailment is measured in doses of milligrams (not pounds or teaspoons).

Having a relatively universal system is important for scientists in order to have a basis for comparison. If, for example, a biologist in France is measuring the response humans have to high doses of caffeine, and he tests his subjects by giving his participants 5000 mg of caffeine, as scientists we immediately have an understanding of the amount of treatment the French subjects are receiving. If the treatment was given in another type of measurement (i.e., "*a pinch*" or "*a lot*" of the dose), there is no way to know how much they are actually receiving.

This is important not only for the interpretation of scientific studies but also for the practice of **replication**. Scientists repeat other researchers' work to see if their results coincide. The results of replication indicate how well a particular researcher performed her experiment and how much emphasis should be placed on a given scientific result. This is why we hear that groundbreaking studies are contradicted by another study later in time. Be aware of this particularly in the media; their standards for what is "proven" by science is often not supported by the data or the scientific community at large. For a particular study to be given full endorsement by other scientists, it must withstand the scrutiny of many other researchers all over the world. Replication is the reason that the next new "wonder drug" may not be available for 5 years: it must be tested and retested, using the same procedures, to be sure the primary results hold up to the practice of replication.

The standard units of measurement are as follows: Length is measured in **meters**, mass in **grams**, volume in **liters**, and temperature in degrees **Celsius** (Table 3.1). *These basic units (meter, liter, gram, degree Celsius) are the starting points of the metric system.* What makes this system so user-friendly is that it is based on a decimal system (divisions of 10). As an example a millimeter is *1/1000th* of a meter and a kilometer is *1000* meters. To use this system, remember the following: (1) calculate a number smaller than the unit indicated by using a subunit prefix and moving the decimal place to the right, and (2) calculate one larger than the unit indicated by using a superunit prefix and moving the decimal place to the left (Table 3.2). If this is the first time you have studied the metric system, it may seem complicated, but it is actually much easier to learn than the customary U.S. system because it is entirely based on the decimal plan, that is, using divisions of 10.

TABLE 3.1 Metric Units of Measure and Their Equivalents

Measurement	Metric Unit	Equivalent
Length	Meter (m)	1 m = 3.280 feet 1 in = 2.5 cm 1.61 km = 1 mile 1 km = 0.62 mile
Volume	Liter (L)	1 L = 1.057 quarts
Mass	Gram (g)	1 g = .0022 pound (lb) 1 kg = 2.2 lbs
Temperature	Degree Celsius (°C)	See next page for formulas

TABLE 3.2 Divisions of Metric Units

Prefix	Symbol	Written Value	Number Value	Examples
Larger than the standard unit				
Mega-	M	One million	$1,000,000 = 10^6$	1 Mm = 1,000,000 m
Kilo-	K	One thousand	$1000 = 10^3$	1 kg = 1000 g
Hecto-	H	One hundred	$100 = 10^2$	1 hg = 100 g
Deka-	Da	Ten	$10 = 10^1$	1 daL = 10 L
Smaller that the standard unit				
Deci-	d	One-tenth	$.1 = 10^{-1}$	10 dL = 1 L
Centi-	c	One-hundredth	$.01 = 10^{-2}$	100 cm = 1 m
Milli-	m	One-thousandth	$.001 = 10^{-3}$	1000 mg = 1 g
Micro-	u	One-millionth	$.000001 = 10^{-6}$	1,000,000 um = 1 m

Conversions are easy using the metric system compared to the traditional U.S. weights and measures, where nearly every conversion requires multiplication or division by a different quantity. To convert a measurement in meters to centimeters, or a quantity of grams to kilograms, or liters to centiliters, simply move the decimal place.

For example:

$$1.52 \text{ meters} = 152 \text{ centimeters}$$

(Because there are 100 centimeters in 1 meter, and because we are converting from a larger unit to a smaller one, move the decimal two places to the right.)

$$500 \text{ milliliters} = .5 \text{ liters}$$

(Because there are 1000 milliliters in 1 liter, and because we are converting from a smaller unit to a larger one, the decimal place is moved three places to the left.)

What would you do if you knew your weight in pounds, but needed it in grams or kilograms? You have to perform another conversion. Table 3.1 tells us that 1 kg is equal to 2.2 pounds (lb). To convert pounds into kilograms, we divide our weight (lbs) by .0022 lbs.

For example if you weighed 200 pounds, you would weigh 91 kg:

$$200 \text{ lb} \div 2.2 \text{ lb/kg} = 91 \text{ kg}$$

Temperature Scales

The metric system utilizes the Celsius scale to measure temperature. Using the Celsius scale is simple because the freezing point of water is 0°C and the boiling point of water is 100°C. Compare this with 32°F and 212°F on the Fahrenheit scale. Converting between these two scales is not as easy as moving a decimal place. Celsius and Fahrenheit temperatures can be converted quickly by reading from one scale across to the other on a thermometer. Conversion is also achieved by using the following simple formulas:

$$°F = (°C \times 1.8) + 32$$
$$°C = \frac{(°F - 32)}{1.8}$$

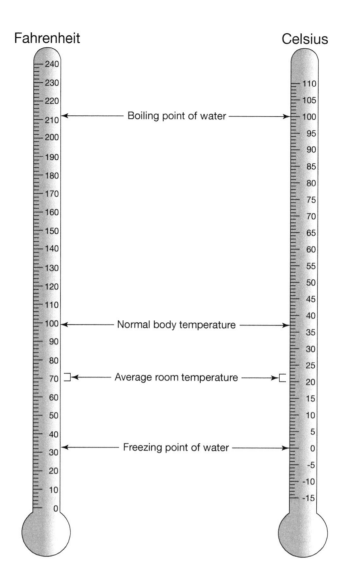

The Scientific Method

The Scientific Method is nothing more than an ordered system of testing observations. It is a universally accepted way of looking at the world and attempting to give meaning to what we see through research and experimentation. It always begins with the simple observation of something in the environment—and then *wondering why* something is the way it is. Scientists use all sorts of tools to enhance or enable our ability to see some phenomenon. Looking through a microscope or telescope, for example, allows us to see what is commonly invisible to the unaided eye.

Once a phenomenon has been observed, and some background has been researched, scientists are able to form a **hypothesis**. A hypothesis is a logical statement that answers a question or explains an observation. A hypothesis is not a fact, but is meant to be tested, challenged, and refined as a result of experience. A testable prediction is made and a controlled experiment is designed. The hypothesis is tested, the data are collected and analyzed, then a conclusion is made. Remember a hypothesis is not proven. It is accepted or rejected.

Lab 3: Pre-Lab Questions

Name: _____

Read through **all** of the lab handout and answer these questions **before** coming to class.

1. Which system of measurement is the most widely adopted?
   a. Metric
   b. Standard
   c. Scientific
   d. English

2. Which is **not** a metric unit of measurement?
   a. Liter
   b. Gram
   c. Cup
   d. Meter

3. Which of the following is a logical statement that answers a question or explains an observation?
   a. Theory
   b. Hypothesis
   c. Result
   d. Idea

4. What is the process of scientists repeating other researchers work to see if their results coincide called?
   a. Repetition
   b. Re-runs
   c. Objection
   d. Replication

5. Which of the following is the base metric unit of measurement for length?
   a. Meter
   b. Celsius
   c. Liter
   d. Gram

6. Which of the following is the base metric unit of measurement for mass?
   a. Meter
   b. Celsius
   c. Liter
   d. Gram

7. Which of the following is the base metric unit of measurement for volume?
   a. Meter
   b. Celsius
   c. Liter
   d. Gram

8. The metric system of measurement is user friendly because it is based on a _____ system.
   a. Obvious
   b. Counting
   c. Decimal
   d. Linear

9. A hypothesis must account for all available information and be _____?
   a. Testable
   b. Intelligent
   c. Scientific
   d. Correct

10. Which of the following is the relationship between the mass and volume of an object?
    a. Grams
    b. Gravity
    c. Acceleration
    d. Density

Procedure

For the following experiments, fill in the corresponding charts and answer the questions on the answer sheet.

Part 1—Height : Weight Ratio

In this exercise, you will use the scientific method to study the relationship of various characteristics of the human body to one another. In the early 1800s there was actually a theory in forensic science that criminals had certain ratios between their head diameter, arm length, height, and weight. People were even convicted of crimes according to their head size. You have probably observed a relationship between height and weight. Tall people usually weigh more than short people. But what is the specific relationship? We will construct a hypothesis for the relationship between height and weight and test it.

- ***Hypothesis:*** *Height in centimeters divided by weight in kilograms equals 2 cm/kg.*

    a. Test the hypothesis by determining your weight (***in kilograms***) and your height (***in centimeters***). Record your data on the answer sheet.
    b. Answer the questions found in Part 1 of your answer sheet. If your height : weight ratio is between 1.9 and 2.1 cm/kg then you support the hypothesis otherwise you refute the hypothesis.

Part 2—Arm Length : Height Ratio

You have probably observed that tall people have long arms, whereas shorter people have shorter arms. Just like in Part 1, let's construct a hypothesis.

- ***Hypothesis:*** *The length of a person's arm in centimeters is equal to 0.4 of his or her height in centimeters.*

    a. Using the height you already measured, determine what we can expect the length of the arm to be by multiplying your height in centimeters by 0.4. Enter this number in the table under "Expected Arm Length" on the Answer Sheet Part 2.
    b. Measure arm length in one of your group members. Record this arm length under "Actual Arm Length".
    c. Subtract the "Actual" value from the "Expected" value. Record this difference under "Deviation from Expected". If the deviation of your arm length is less or equal to 0.5 cm then you support the hypothesis otherwise you refute the hypothesis.
    d. Answer the questions listed under Part 2.

Part 3—Hand Temperature

Normal internal body temperature is 98.6°F in humans. Is the internal body temperature the same as hand temperature?

a. On the answer sheet construct a hypothesis correlating hand temperature and body temperature (refer to the examples hypotheses above and in the introduction for help). Make sure that your hypothesis is testable by experimentation and accounts for any information you know about body temperature differences.

b. Use a thermometer to determine your hand temperature. Hold the bulb of a thermometer in you hand for 2 minutes.

c. Answer the questions under Part 3.

Answer Sheet

Lab 3: Metric System and Scientific Method

Name: _____   Section: _____

Part I – Height : Weight Ratio

a. Fill in the following chart:

Height (cm)	Weight (kg)	Height ÷ Weight (Actual)	Height ÷ Weight (Expected)
			2 cm/kg

b. Answer the following question:

Is height in centimeters equal to two times weight in kilograms? If your height : weight ratio is between 1.9 and 2.1 cm/kg then you support the hypothesis otherwise you refute the hypothesis.

Part II – Arm Length : Height

a. Fill in the following chart:

Height (cm)	Arm Length Expected (cm) * To get this number take the height and times it by 0.4	Actual Arm Length (cm)	Deviation from expected (Actual − Expected)

b. Answer the following questions:

Describe your procedures for taking the arm length measurement. Was a standard for how to measure arm length given in the procedures? Why is it important, in scientific terms, to have measurements like these standardized?

Did your results support or refute the hypothesis that arm length in centimeters is 0.4 of height in centimeters? If the deviation of your arm length is less or equal to 0.5 cm then you support the hypothesis otherwise you refute the hypothesis. Do you think any of your classmates found alternate results? Why or why not?

Part III – Hand Temperature

State your hypothesis about the relationship between internal body temperature and hand temperature.

What was your hand temperature in °C? Normal internal body temperature is 98.6°F. Convert this to °C below too.

Was your hypothesis supported or rejected? Describe other factors that may influence hand temperature.

Part IV – Fill in the charts

Be sure to include units (g, cm, etc.) in your answer

Item	Width in meters	Width in centimeters
A piece of paper		
Your lab workstation		

Item	Mass in grams	Mass in kilograms
You!		

Part V – Answer questions below

1. Lead melts at 620°F. Would a furnace that heats to 200°C melt the lead? Why or why not?

2. Your instructor loves ice cream! Would he/she rather have 1 kg of Rocky Road or 10,000 mg? Express both in grams.

3. The pharmacist has 1 quart of liquid Vitamin D and a patient brings in a prescription for a 20 mL bottle. How many quarts will the pharmacist put into a bottle to dispense to the patient?

4. The orthopedic surgeon needs to cut off 15 inches of the femoral head to attach a new metal replacement. How many millimeters of the bone will he remove?

5. Place the decimal point in its correct place for the following statements:

Ex: Toby lost 200 kg last week	Toby lost 2.00 kg last week
Susan is 13,525 cm tall	
The heart pumps 754.1 mL of fluid	
In a myelinated neuron, an action potential travels at 10 m/s	

Laboratory 4

Chemistry

Objectives

- Determine, label, and draw the number of protons, neutrons, and electrons for any given atom
- Describe and diagram ionic and covalent bonds
- Be able to identify functional groups
- Define organic compound
- Determine if a given solution is an acidic or basic
- Build molecular models

Introduction

Why Do We Need Chemistry *to Understand* Biology?

As you are surely aware biology is the study of living things, organisms. When we deduce life to its most basic foundation, we are looking at cells. When we look a little further, we see that a cell is made of tiny bits of matter. The cell membrane, the part of a cell that contains it and regulates what enters and what leaves, is made of only a handful of types of atoms, bonded together into larger molecules. Membranes are made of millions of lipid molecules (a type of fat) that are comprised of carbon, hydrogen, and oxygen atoms bound to a group of other atoms that happen to include a phosphate atom, as well as a few hydrogen and oxygen atoms.

Possessing a working understanding of some chemistry basics is as critical to our knowledge of biology as knowing that cells are the basic building blocks of life, or that the sun's energy is used by plants to make food. Chemistry is the foundation for life and living things. It is the interaction between and among atoms that make up the chemical reactions we call life.

Atoms

The smallest unit of matter (including living matter) that retains the properties of an element is known as an **atom**. There are 92 different atoms found in nature and several others that have been created in laboratories, but disintegrate quickly. Matter that consists of exclusively of one kind of atom is known as an **element**. Atoms can be combined in many ways to form countless different kinds of molecules. Each kind of molecule has a specific arrangement of atoms within its structure. Matter that is composed of only one kind of molecule are called **compounds**. Each

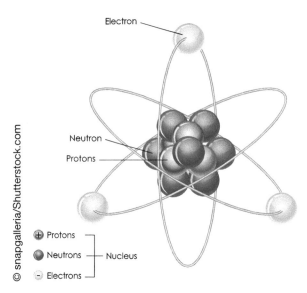

FIGURE 4.1 Structure of an atom.

kind of atom has a specific number and arrangement of parts that differs from the number and arrangement of parts in other kinds of atoms. The specific structure of an atom determines the kinds of atoms that it can bond with to form larger molecules. Each atom consists of three types of subatomic particles: **protons, neutrons,** and **electrons**. The positively charged protons and uncharged neutrons are located in a central area called the **nucleus** and the negatively charged electrons move around outside the nucleus in specific regions known as electron **shells** (Figure 4.1).

The periodic table of the elements arranges elements in order of increasing complexity and according to how they react chemically. It identifies the number of protons, neutrons, and electrons in atoms of each element. Figure 4.2 demonstrates how to obtain this information about a carbon atom from the periodic table. Each element has a one or two letter symbol; Carbon's is *C*, Sodium's is *Na*, Chlorine's is *Cl,* Gold's is *Au,* etc.

Atomic number: Determining the number of protons and electrons

The number in each box that is a whole number (no decimal places) is the **atomic number.** The atomic number tells us how many protons (positively charged particles) there are in an atom's nucleus. It is the identifying characteristic of an atom; it does not change. Carbon, for example, will *always* be number 6 and will *always*, therefore, have 6 protons in its nucleus. Hydrogen (number 1) has one proton—*always*. The number of protons in an atom affects the number of electrons that exist in that atom's outer orbitals because *atoms are neutral*. Therefore, the total number of positive particles (protons) must *exactly match* the number of negative ones (electrons).

Mass number: Determining the number of neutrons

To determine the number of neutrons, the **mass number—a.k.a. atomic mass—**is used. Even though there are three types of subatomic particles in an atom, electrons have almost no mass. The atomic mass, then, is essentially the mass of the nucleus because the nucleus contains the particles that do have mass: the protons and neutrons. The mass of a proton and a neutron are the same: 1 atomic mass unit (amu). *The total atomic mass is the sum of the number of protons*

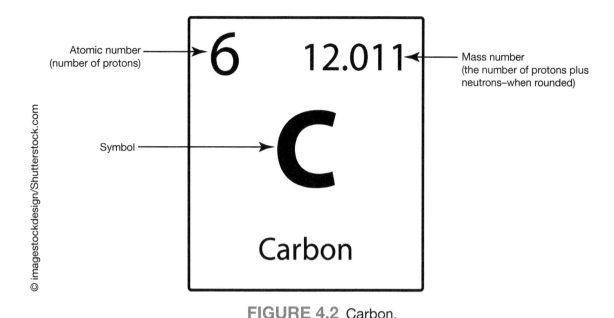

FIGURE 4.2 Carbon.

and neutrons in the atom. You can determine the number of neutrons in a carbon atom, as an example, by subtracting the atomic number 6 from the mass number 12 (rounded off); you will find that there are 6 neutrons in a carbon atom (12 − 6 = 6). To determine the number of protons, electrons, or neutrons in an atom, use the periodic table and the following equations:

1. Atomic number = number of protons
2. Atomic number = number of electrons (in a neutrally charged atom)
3. Mass number (rounded off) − atomic number = number of neutrons (in a typical atom)

Electron arrangement

It is true that opposite charges attract and like charges repel. This is the reason negatively charged electrons (rotating around the nucleus) are held near the positively charged center of the atom (full of protons: carrying a positive charge). Their movement prevents them from being pulled into the nucleus. Considering electrons are negative, they will push away from each other and move around the nucleus in specific regions called **orbitals**. The distance of the orbitals from the nucleus is determined by the amount of energy the electron possesses. Electrons that are furthest from the nucleus have the greatest energy. Remember that electrons will move as far away from each other as possible; the more room available at a particular level, the more orbitals and, therefore, electrons. The first shell or energy level can hold a maximum of 2 electrons. The next shell can hold 8. The lowest energy shell must always fill first.

Ionic bonds

Some kinds of atoms have such a strong attraction for electrons that they steal electrons from other atoms that have loosely held electrons. The atom who "steals" the electron has, then, gained one and is now a negatively charged ion because it has an extra electron (recall that an ion is formed when the number of protons and electrons do not match). The atom that has given up an electron is now a positive ion; it has one more proton than it has electrons. The formation of oppositely charged ions means that they will now have an electrochemical attraction and they will develop an ionic bond. The old statement is true: opposites do attract, especially in chemistry. Like charges (two of the same, ++ or −−) will repel each other.

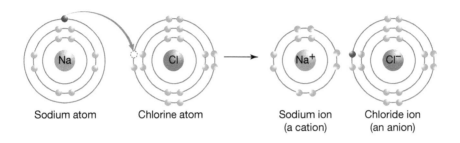

Sodium atom   Chlorine atom   Sodium ion (a cation)   Chloride ion (an anion)

Covalent bonds

A second kind of bond that holds atoms together to form molecules is known as a **covalent bond**. In covalent bonds the electrons are not actually transferred from one atom to another, as in the formation of ions and ionic bonds, but are shared by two or more atoms. Each pair of electrons that is shared is the equivalent of one covalent bond. Chemists typically diagram molecules by using a line between atoms to represent a single covalent bond. As the diagram below indicates, a single oxygen atom is sharing 2 electrons with 2 different hydrogen atoms and each of the hydrogen atoms is sharing an electron with the same oxygen. Sometimes two atoms may share more than one pair of electrons, creating a double bond. An example of this is carbon dioxide. The molecule has the following structure: O=C=O. This shows that a carbon atom is sharing two electrons with one oxygen atom and two electrons with another oxygen atom. Each oxygen atom shares two electrons with the same carbon atom.

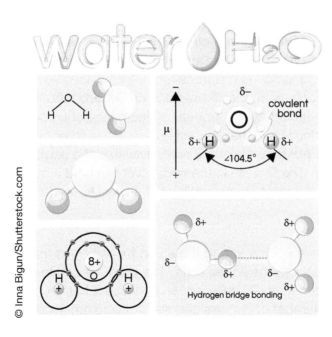

© Inna Bigun/Shutterstock.com

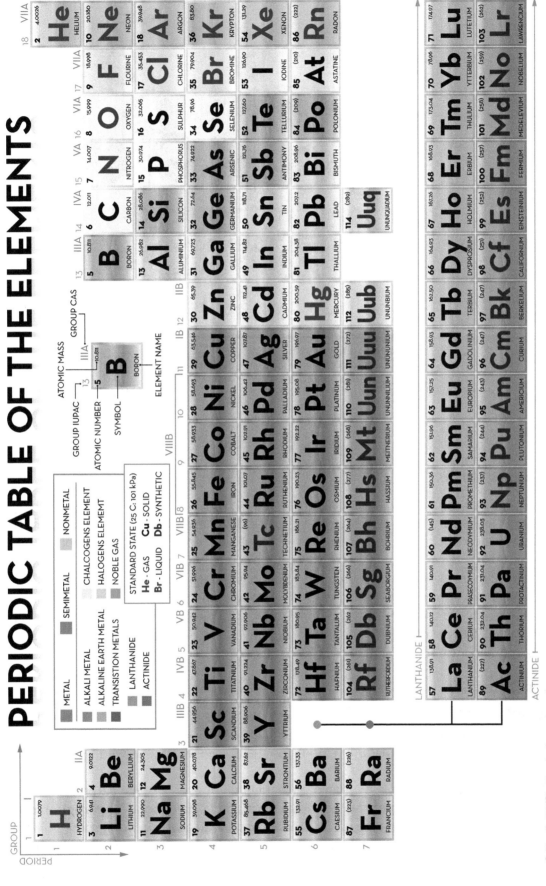

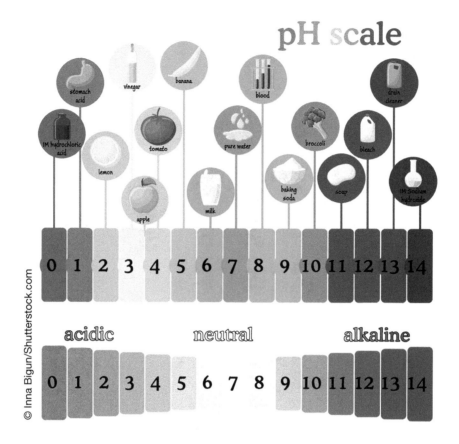

Acids, Bases, and pH

When water ionizes, it releases an equal number of both hydrogen ions (H+) and hydroxyl ions (OH−). Acids are molecules that release hydrogen ions when dissolved in water. A hydrogen ion is a hydrogen atom that has lost its electron (H+). Substances that remove hydrogen ions from a solution are known as bases (OH−). If there are more H+ than OH−, then the solution is an acid, if there is more OH− than H+ then the solution is a base.

The pH scale is a measure of the number of hydrogen ions present in a solution. It is used to indicate the acidity or alkalinity of a solution. A pH of 7 is neutral. The pH of lemon juice is 2, which is acidic. The pH of baking soda is 12, which is alkaline or basic. Refer to the pH scale above for more pH values of common substances.

Organic Molecules

Organic molecules have one characteristic in common; **they all contain carbon atoms**. Carbon is a very versatile atom for creating molecules because it has four electrons in its second electron orbital. These four electrons in a carbon atom can make covalent bonds with four other different atoms.

Hydrocarbons are the simplest organic molecules composed of only carbon and hydrogen atoms. Since neither carbon nor hydrogen are polar, hydrocarbons are **hydrophobic** (water-hating) molecules. Fat is an example of a hydrophobic molecule with long hydrocarbon fatty acid tails that don't mix with water. Methane is the simplest hydrocarbon with one carbon and 4 hydrogen atoms that covalently bond together as shown.

Methane (CH₄)

Simple hydrocarbons like methane can be modified by replacing hydrogens atoms with one or more polar **functional groups** that make the molecules **hydrophilic** (water loving) and give the molecules specific chemical properties. Hydrophilic molecules readily dissolve in water. Sugar is an example of a hydrophilic molecule that we use to sweeten watery drinks like soda pop.

Functional Groups

Name	Formula	Found in	Chemical Characteristics
Hydroxyl (−OH)		Lipids, Proteins, Carbohydrates, Nucleic Acids, Alcohol	Polar, Hydrophilic, Forms Hydrogen bonds
Carboxyl (−COOH)		Lipids, Proteins	Polar, − charge, Hydrophilic, Forms Ionic Bonds
Amine (−NH₂)		Proteins, Nucleic acids	Polar, + charge, Hydrophilic, Forms Ionic Bonds
Phosphate (−PO₄)		Nucleic Acids, Phospholipids	Polar, − charge, Hydrophilic, Forms Ionic Bonds

Our bodies are mostly made up of four types of organic molecules—**Carbohydrates, Lipids, Proteins, and Nucleic Acids**. These organic molecules are usually very large and consist of a long carbon backbone with functional groups attached by covalent bonds.

Each of these organic molecules is built of from single units called **monomers**. These monomers are linked together one by one to form large **polymers** by dehydration reactions.

Dehydration reactions occur between a hydrogen (H) hanging off of one monomer and a hydroxyl functional group (OH) on another. When the two monomers react the hydroxyl and hydrogen are lost as water (H_2O) as shown.

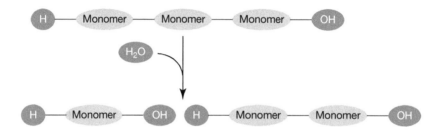

Hydrolysis (literally water splitting) is the reverse of the condensation reaction that breaks down polymers back into their monomer constituents. Hydrolysis splits monomers by adding back the water that was removed in the condensation reaction as shown below.

Our bodies digest the food we eat through hydrolysis reactions that are carried out by specific enzyme proteins in our digestive tract. Our bodies then use these free monomers from food to create the polymers that make up our body. In this way you really are what you eat because most of the polymers that make up your body are made of monomers from the meats, fruits, and vegetables that you eat.

Carbohydrates are made of monomers called **monosaccharides**. Glucose ($C_6H_{12}O_6$) is the most prevalent type of monosaccharide. Two glucose monomers are linked together to form **disaccharides** like lactose (milk sugar) and sucrose (table sugar). Hundreds and thousands of glucose monomers linked together form **polysaccharides** like glycogen (for sugar storage in animals), starch (for sugar storage in plants), and cellulose (for building plant cell walls).

Lipids are made of monomers called **fatty acids and glycerol**. One glycerol molecule reacts with three fatty acids to form fat. Also, one glycerol can react with two fatty acids and a phosphate functional group to form the phospholipids that make up the cell membrane.

Proteins (or polypeptides) are made of monomers called **amino acids**. The **1° structure** of proteins is the sequence of the chain of amino acids linked together by peptide bonds. Parts of the chain then spontaneously fold into two types of domains called the **alpha helix** and **pleated sheet** making up the **2° structure**. These domains then wind around each other to form the globular 3-D **3° structure**. The protein is only functional after the tertiary structure has formed. With some proteins, such as hemoglobin, several of them join together and form the **4° structure**.

Nucleic acids, DNA and RNA, are made of monomers called **nucleotides**. IN DNA there are four type of nucleotides each of which has a specific base or DNA letter of A,T,G, or C. Nucleotides form long chains that wind around each other making a double helix in DNA. In this laboratory activity we will investigate carbohydrates, lipids, proteins, and nucleic acids by investigating functional groups, creating 3-D molecular models, and by learning about chemical formulas.

Lab 4: Pre-Lab Questions

Name: _____

Read through **all** of the lab handout and answer these questions **before** coming to class.

1. An atom consists of
   a. Protons
   b. Neutrons
   c. Electrons
   d. All of the above

2. A bond that is formed between oppositely charged ions (due to electron 'stealing', NOT sharing) is called which of the following?
   a. Ionic bond
   b. Covalent bond
   c. Hydrogen bond
   d. Electron bond

3. Protons and Neutrons are found in which part of the atom?
   a. Orbitals
   b. Nucleus
   c. Space between the nucleus and orbitals
   d. Both the nucleus and orbitals

4. The number of protons in an atom is the?
   a. Atomic mass
   b. Atomic number
   c. Atomic size
   d. Atomic symbol

5. The first electron orbital holds _____ electrons and the second and higher orbitals hold _____ electrons.
   a. 4; 8
   b. 2; 4
   c. 2; 8
   d. 1; 4

6. The hydrogen and oxygen atoms in a water molecule are held together by which type of bond?
   a. Ionic bond
   b. Covalent bond
   c. Hydrogen bond
   d. Electron bond

7. All living organisms contain which of these organic molecules
   a. Lipids
   b. Carbohydrates
   c. Proteins
   d. All of the above

8. Carbon is a versatile atom because it can make _____ covalent bonds.
   a. 1
   b. 2
   c. 3
   d. 4

9. Hydrocarbons contain which elements?
   a. Carbon and Hydrogen
   b. Carbon and Nitrogen
   c. Hydrogen and Nitrogen
   d. Carbon and Sulfur

10. Which type of reaction makes biological organic molecules bigger?
    a. Dehydration reaction
    b. Hydrolysis reaction
    c. Hydrocarbon reaction

11. Which of these is **not** one of the 4 types of atoms most commonly found in organic molecules?
    a. Hydrogen
    b. Carbon
    c. Magnesium
    d. Oxygen

12. Which of the following is **not** a functional group found in organic molecules?
    a. Phosphate
    b. Carboxyl
    c. Organoxyl
    d. Hydroxyl

13. Which of the following is the class of biological molecules that includes fat?
    a. Lipids
    b. Proteins
    c. Nucleic acids
    d. Carbohydrates

14. Which of the following is the class of biological molecules that includes DNA?
    a. Lipids
    b. Proteins
    c. Nucleic acids
    d. Carbohydrates

Procedure

For the following exercises and experiments, fill in the corresponding charts and answer the questions on the attached answer sheet. Make sure that you read the lab and the section in your textbook covering basic chemistry before you begin.

Part I – Atoms and Electron Arrangement

(a) In this exercise, you will use the periodic table of elements to determine the number of protons, electrons, and neutrons in atoms of the listed elements and fill in the blank spaces on the answer sheet. The first one has been done as an example. (b) You will diagram electron arrangements in an atom.

Part II – Ionic Bonds

(a) In this exercise, you will use the periodic table of elements to determine the number of protons and electrons that comprise the following atoms, ions, and molecules. The first two have been done as an example. Fill in the chart located on the answer sheet.

Part III – Acids, Bases, and pH ***Be careful and do not spill the solutions or get them on your skin, some are strong acids and bases. You MUST wear gloves when handling these chemicals. Goggles and aprons are also available. ***

a. In this exercise you will be given pH values and asked to determine their acid and base concentration compared to other pH values. Circle the correct answers on your answer key.

b. Unknown and Household pH Exercise – Many kinds of materials change color as the pH of a solution changes. We can make use of this property to determine the pH of an unknown solution. There are 5 unknown solutions available for you to test the pH of. They are labeled A to E. You will also be testing 5 household items pH.

    i. Put 5 strips of pH paper on a paper towel. Place one drop of the unknown solution onto one end of a piece of pH paper. Compare the color of the paper with the information on the dispenser chart and record the pH value in the answer sheets table.

    ii. Place 5 drops of solution A into a well on a spot plate and add two drops of bromothymol blue.

    iii. Record the color of the solution in the table on the answer sheet

    iv. Place 5 drops of solution A into another well and add two drops of phenolphthalein. Record the color of the solution in the table on the answer sheet

    v. Repeat the steps with solutions B-E and a selection of household products

    vi. Dispose of each solution into the sink. Rinse off the spot plate with tap water and place it into the "dirty dish bin"

c. Answer the questions listed

Part IV – Organic Molecules

(a) Building Models of Organic Monomers

In this activity, you will use the ball and stick molecular model kits to construct the organic monomers. You will need to get signed off that you have completed them correctly.

(b) The Elements of Organic Molecules

In this exercise, you will use the periodic table of elements to determine the atomic number, atomic symbol, number of electrons in outermost (valence) shell and the number of bonds these elements will make.

(c) Functional Groups in Organic Molecules

In this exercise, you will determine the chemical formula of some common organic molecules. You will also identify functional groups in these organic molecules.

Clean-up – Place glassware into the labeled dishpan. Throw away the used pH papers and any other paper items into the trashcan.

Answer Sheet

Lab 4: Chemistry

Name: _____ **Section:** _____

Directions
- Fill in the following fields as completely and accurately as possible.

Part I – Atoms and Electron Arrangement

(a) Fill in the following chart:

Element	Symbol	Protons	Neutrons	Electrons
Carbon	C	6	6	6
Hydrogen				
Nitrogen				
Oxygen				
	Na			
	Mg			

(b) Draw the following:

Diagram an atom of the element with the atomic number 11. Include the number and position of the protons, neutrons, and electrons. You do not need to draw each electron, neutron, and proton individually. Just write the number of each particle in the orbital/nucleus. Also name the atom.

Part II – Ionic Bonds

(a) Fill in the following chart:

Atom/Ion/Molecule	Number of protons	Number of electrons
Na atom	11	11
Na$^+$ ion	11	10
Ca atom		
Ca^{2+} ion		
O atom		
O^{2-} ion		
CaO molecule		

Part III – Acids, Bases, and pH

(a) Circle the correct answers

1. Which of the following pH is more acidic?
    a. pH 1 or pH 4
    b. pH 5 or pH 3

2. Which of the following pH values is more basic?
    a. pH 8 or pH 10
    b. pH 7 or pH 11

3. What number on the pH scale represents a neutral pH?
    a. 14
    b. 7
    c. 1
    d. 5

4. Which of these is **not** a way that we will be using to determine the pH of a solution?
    a. pH paper
    b. Bromothymol blue
    c. Mass number
    d. Phenolphthalein

(b) Fill in the following chart: For A to E and for 5 EVERYDAY CHEMICALS

Solution	pH Paper (write number)	With Bromothymol Blue (write color)	With Phenolphthalein (write color)
A			
B			
C			
D			
E			
Windex			
Lemon Juice			
Mouthwash			
Vinegar			
Soapy Water			

(c) Answer the following questions:

Bromothymol blue changes color when mixed with an acid. What color does it become?

What color is bromothymol blue when it is in a base?

What color would you expect bromothymol blue to be if it were in a neutral solution?

Phenolphthalein changes color to _____ in the presence of base.

What color is phenolphthalein in an acid?

Which indicator solution (bromothymol blue or phenolphthalein) could be considered to be an acid indicator?

Explain how one might use bromothymol blue and phenolphthalein to test the pH of water in a swimming pool. What would be the result if the water in the pool is pH 10?

Part IV

(a) Building Models of Organic Monomers

Use your ball and stick molecular model kit to construct the organic monomers below. Refer back to the diagrams of the monomers (in the introduction) for assistance in building these models. **Have the instructor or instructional aides check your structures and initial below.**

1. Glycerol _____

2. Monosaccharide _____

3. Amino acid _____

(b) The Elements of Organic Molecules

There are four types of atoms that are most commonly found in organic molecules. They are carbon, hydrogen, oxygen, and nitrogen. Others that are frequently found in organic molecules include phosphorus and sulfur. Review the bonding behavior of these atoms by filling out the following chart (use your textbook and the periodic table in the previous chemistry lab for assistance). Remember that the first orbital can have up to two electrons and the next two orbitals can have up to eight electrons each. Each orbital must be filled before the next one can accept electrons.

Atom Name	Atomic Number	Symbol	# of Electrons in the Outermost Orbital	# of Covalents Bonds Possible
Carbon	6	C	4	4
Hydrogen				
Oxygen				
Nitrogen				
Phosphorus				
Sulfur				

(c) Functional Groups in Organic Molecules

On the next page, **circle** and **identify** the functional groups in the following molecules. Some of the molecules will have more than one functional group attached to the carbon skeleton. Also **count** the different types of atoms and **indicate** the chemical formula of the molecules.

Here are a few examples

This molecule has 3 hydroxyl groups.

It also has 3 Carbons (C_3), 8 Hydrogens (H_8), and 3 Oxygens (O_3).

Chemical Formula $C_3H_8O_3$

This molecule has one carboxyl group.

It also has 3 Carbons (C_3), 6 Hydrogens (H_6), and 2 Oxygens (O_2)

Chemical Formula $C_3H_6O_2$

Circle the functional groups and **indicate the chemical formulas** for the following organic monomer molecules on your own. Refer back to the previous page for the chemical formula and structures of the functional groups.

Fatty acid (1 Functional group)
Chemical Formula

Glycerol (3 Functional groups)
Chemical Formula

Amino acid (2 Functional groups)
Chemical Formula

Nucleotide (4 Functional groups)
Chemical Formula

Monosaccharide (5 Functional groups)
Chemical Formula

Part V

This portion of the Lab is designed to help give you a working knowledge of atoms bonding properties.

Below two or more atoms are listed. Tell me each atom's atomic number, how many protons and how many electrons each atom has, draw the Bohr's Planetary Model for each atom, then tell me whether an IONIC, COVALENT, BOTH or NO BOND is formed. Keep in mind two electrons in 1st shell and eight in the following shells. Atoms want a full valence shell.

1.   H   H

Atomic #   _____   _____
Protons   _____   _____
Electrons   _____   _____

Draw Models:

Circle: Ionic Bond   Covalent Bond   Both   No Bond would form

Can you guess the name of the molecule: _____

2.   O   O

Atomic #   _____   _____
Protons   _____   _____
Electrons   _____   _____

Draw Models:

Circle: Ionic Bond   Covalent Bond   Both   No Bond would form

Can you guess the name of the molecule: _____

3.   H   Cl

Atomic #   _____   _____
Protons   _____   _____
Electrons   _____   _____

Draw Models:

Circle: Ionic Bond   Covalent Bond   Both   No Bond would form

Can you guess the name of the molecule: _____

4.   Na   O   H

Atomic #   _____   _____   _____
Protons   _____   _____   _____
Electrons   _____   _____   _____

Draw Models:

Circle: Ionic Bond   Covalent Bond   Both   No Bond would form

Can you guess the name of the molecule: _____

5.   Mg   S

Atomic #   _____   _____
Protons   _____   _____
Electrons   _____   _____

Draw Models:

Circle: Ionic Bond   Covalent Bond   Both   No Bond would form

Can you guess the name of the molecule: _____

6.   Li   F

Atomic #   _____   _____
Protons   _____   _____
Electrons   _____   _____

Draw Models:

Circle: Ionic Bond   Covalent Bond   Both   No Bond would form

Can you guess the name of the molecule: _____

Laboratory 5

Microscopes and Cells

Objectives

- Demonstrate the ability to use a compound microscope
- Correctly set-up and focus the microscope
- Properly handle, clean, and store the microscope
- Correctly use all lenses of the microscope
- Record microscopic observations
- Be able to distinguish various components of living cells using a light microscope
- Understand and adjust the contrast on a microscope
- Make wet mounts

Introduction

Because this is a Biology class, we will use the microscope to look at cells and many microorganisms. It is an important tool for biologist. Therefore, it is essential to become proficient with its use. The microscope you will be using is an expensive instrument that needs to be treated with respect. You are responsible to properly care for, clean and store the microscope you are using. If you find that the microscope is not working properly, notify your instructor or instructional aide immediately. They will fill out a repair card so that it may be fixed. Do not try to fix the microscope yourself.

Other students will use the same microscope in other lab classes. Be sure to remove slides and clean debris and stain from the stage before storing. No one likes to start lab with a dirty microscope.

Components of the Microscope

Look at the picture of a microscope in Figure 5.1. Locate the following:

1. **Arm:** This is the vertical support of the microscope and the major structural element. When moving or carrying the microscope the arm is grasped firmly while supporting the base with the other hand.

2. **Base:** The base of the microscope supports the arm and provides stability for the optical system. The on/off switch is on the base and in many microscopes the light source is built into the base. The **rheostat** controls the brightness of the light.

3. **Binocular eyepiece tube:** This is a complex assembly that holds the ocular lenses. It is designed to allow adjustment of the distance between the oculars and the focus of one of them. The eyepiece tube is held in position on top of the arm by a setscrew. Loosening the setscrew could result in the eyepiece tube falling and being damaged.

4. **Revolving nosepiece** and **objectives:** The revolving nosepiece is attached to the eyepiece tube and holds the objective lenses. The nosepiece is designed to allow it to be rotated so that objectives of different magnification are clicked into place. There are threads on the rim of the nosepiece that make it easy to grasp and rotate. **Do not** hold the objectives when rotating the nosepiece to a different magnification.

5. The **stage** is the flat surface where the slides are placed for viewing.

6. Focus controls raise and lower the stage so that the distance can be varied between the specimen and the objective. The larger knob is called the **coarse focus** and the smaller knob is called the **fine focus**. The stage moves a great distance with the coarse focus and a small distance with the fine focus. Because the fine focus has a limited range of movement, sometimes the fine focus will not adjust the focus of the specimen adequately. To fix this, you need to rotate the 10× objective into position, and then rotate the course focus to the middle of its range. Next use the fine focus to bring the specimen into focus. This will take

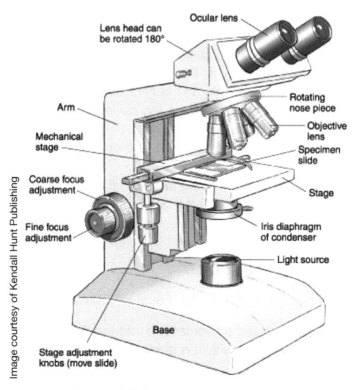

FIGURE 5.1 Compound Microscope.

an extremely small adjustment. After completing these steps, the fine focus should be useable. When using the coarse focus of the microscope, **never** focus "up" while looking through the oculars. A collision could occur between the slide and the objective, which could damage or break both. Also never, use the coarse focus with high power.

7. The **mechanical stage assembly**: This assembly allows you to move the slide side-to-side and front to back to position the proper area of the specimen under the objective for viewing. This assembly consists of two parts: (1) the **moveable spring slide clamp** that hold the slide in place, and (2) two **stage motion control knobs**, one that moves the slide front and back and one knob that moves the slide side to side.

8. **Condenser mounting bracket**: This bracket holds the **condenser** and **iris diaphragm**. The condenser focuses light onto the specimen. The iris diaphragm regulates the amount of light being passed through the condenser. The **condenser focus knob** adjusts the position of the condenser relative to the stage.

The Lens System

The compound microscope uses a four-lens system: (1) the lens associated with the light source, (2) the condenser lens, (3) the objective lenses, and (4) the ocular lenses. It uses a 2-lens or **compound** system to magnify the specimen. The initial magnification occurs in the objective lens and second magnification occurs in the ocular lens. The **ocular lens** magnifies the image formed by the objective lens ten times (10×). To determine the **total magnification** of the specimen, multiply the magnification of the objective by ten. For example, if using the 10× objective, then the specimen would be magnified one hundred times (100×).

Compound Microscope Magnification Table		
Eye Piece Power	Objective Lens Power	Total Magnification
10×	4× – Scanning	40×
10×	10× – Low Power	100×
10×	40× – High Power	400×
10×	100× – Oil Immersion	1000×

Most microscopes have at least three objective lenses attached to the nosepiece. The objective lenses are identified as a **scanning lens (4×)**, **low-power objective (10×)**, **high-dry objective (40×)**, and an **oil immersion objective (100×)**. Each objective is also designated by different terms. These terms give either the **linear magnification** or the **focal length**. The focal length is equal to or greater than the **working distance** between the focused specimen and the tip of the objective. For example, the low-power objective is also called the 10× or 16 mm objective, the high dry is also called the 40× or 4 mm objective and the oil immersion is called the 100× or 1.8 mm objective. As magnification increases the working distance decreases, and the light entering the tip of the lens also decreases. As magnification increases, the **field of view** and the **depth of view** also decrease.

When changing the objective it is usually required to change the position of the iris diaphragm to increase or decrease the amount of light passing through the condenser and entering the objective. The iris diaphragm is also used to adjust the **contrast** of the specimen. As the diaphragm is closed, specimen contrast is increased, but microscopic resolution is decreased. The stage condenser also controls the amount of light that is focused onto the specimen.

We will not be using the oil immersion lens in this class. However, when it is used, the immersion oil fills the space between the specimen and the objective lens. Immersion oil has the same **refractive index** as the glass of the lens and of the slide. Because of this, the loss of light is minimized. Because we will not be using oil immersion, this objective lens should NEVER cross the stage. You will notice that this lens (labeled 100×) is longer than the other three lenses. If you allow this objective to cross the stage, it could hit the slide or stage and break the lens and/or slide. Therefore, do NOT rotate the 100× lens across the stage.

Cells

Cells are the basic units of which all living organisms are composed. Although cells vary in organization, size, and function, all share three structural features: (1) All possess a **plasma membrane** defining the boundary of the living material; (2) all contain a region of **DNA** (deoxyribonucleic acid), which stores genetic information; and (3) all contain **cytoplasm**, everything inside the plasma membrane that is not part of the DNA region. Please review your textbook for a more in depth details.

In the microscope portion of the exercise, you learn about the important principles that affect how we view an image in the microscope. One of the limitations is size and another is image contrast. Contrast is defined as the ability to distinguish an object from its background. One way we can increase contrast is by closing the iris diaphragm and reducing the amount of light that is shown on the image. As you look at the specimens in lab today, experiment with the iris diaphragm, and notice the difference in contrast as you open and close it.

Another way to increase contrast is through differential staining. Some stains have an affinity to various substances of cells. By using differential staining techniques, we are able to distinguish certain cellular components from the cell background. In this lab we will be using two different stains: Lugol's iodine and methylene blue.

Lugol's iodine (I_2KI) is a mixture of iodine and potassium iodide. I_2KI is used to stain starch granules in cells. The reaction of the iodine with starch produces a complex that appears blue or black in color.

Methylene blue is a vital stain. This means that the ability of it to stain cells is based on the cell actually taking up the stain. Methylene blue binds to basic substances (stains them blue) such as nuclei and ribosomes.

Lab 5: Pre-Lab Questions

Name: _____

Read through **all** of the lab handout and answers these questions **before** coming to class.

1. The correct way to carry a microscope is by grasping the _____ firmly and supporting the base with the other hand.
   a. Objective
   b. Iris diaphragm
   c. Arm
   d. Stage

2. The _____ is the flat surface where the slides are placed for viewing.
   a. Objective
   b. Iris diaphragm
   c. Arm
   d. Stage

3. The _____ is used to adjust the contrast of a specimen by changing the amount of light that passes through the condenser.
   a. Objective
   b. Iris diaphragm
   c. Arm
   d. Stage

4. _____ allows us to distinguish certain cellular components from the cell background.
   a. Differential staining
   b. Brightness
   c. Coloration
   d. Objectives

5. Which of these increase the magnification of the specimen?
   a. Mechanical stage assembly
   b. Condenser
   c. Objectives
   d. Focus controls

6. Which of these moves the slide up and down to sharpen the image?
   a. Mechanical stage assembly
   b. Condenser
   c. Objectives
   d. Focus controls

7. To determine total magnification multiply the power of _____ by the power of the objective lens.
   a. the lens associated with the light source
   b. the condenser lens
   c. the objective lenses
   d. the ocular lenses

8. Which part of the eukaryotic cell acts as a barrier that surrounds the cell?
   a. Plasma membrane
   b. Cytoplasm
   c. Nucleus
   d. Chloroplasts

9. _____ is the ability to distinguish one object from its background.
   a. Differential stain
   b. Brightness
   c. Coloration
   d. Contrast

10. What is the smallest unit of living matter that can reproduce?
    a. Atom
    b. Nucleus
    c. Cell
    d. Organism

11. The differential stain _____ is used to label starch grains in plant cells.
    a. Lugol's iodine (I_2KI)
    b. Methylene Blue
    c. Wet mount
    d. Congo Red

12. Which stain will be used on your cheek cells?
    a. Lugol's iodine (I_2KI)
    b. Methylene Blue
    c. Wet mount

Procedure

Using the microscope

Part I – Getting Started

1. Obtain a microscope from the cabinet by grasping the arm firmly and supporting the base with the other hand, and slowing pulling it from the cabinet so the oculars do not bang into the top or sides of the cabinet.

2. Gently place the microscope onto your workbench, with a paper towel under it, unwrap the rubber band around the cord and plug it in to the outlet.

3. Turn on the light source under the stage and raise the brightness to an intermediary level.

4. Rotate the 4× objective into place if it is not already in position. The microscope should always be put away with the 4× objective in place.

5. Open the condenser diaphragm completely by rotating the knob under the stage until the maximum amount of light comes through.

6. Rotate the coarse focus knob until the stage is at the uppermost point. Do this while looking at the stage not while looking through the eye pieces.

Part II – Letter "e" Slide

1. Go to the common workstation and retrieve a letter 'e' slide.

2. Use a piece of lens paper to clean the slide if necessary.

3. Place a slide of the letter "e" on the stage by moving the slide holder to keep it in place. Center the "e" in the center of the light.

4. Adjust the distance between the two oculars, while looking through the microscope, to accommodate to the distance between your eyes.

5. Some microscopes have one ocular that is fixed and one that is adjustable with a diopter ring. This permits the viewer to compensate for the differences in the vision of each eye. To make this adjustment, keep the eye that looks through the ocular with the fixed focus open while Closing the other eye. Focus the microscope until you see a sharp image. Now look through the eye that was previously closed and adjust the ocular with the diopter ring until the image is sharp. The microscope is now adjusted to the vision of each eye. If you are near-sighted or far-sighted, do not use your glasses with the microscope. The microscope adjusts for faulty vision. If you have astigmatism, the microscope cannot correct for it.

6. Look through the ocular while slowly lowering the stage by rotating the coarse focus knob downwards. **Never rotate the coarse focus upwards while looking through the oculars**. Bring the "e" into focus (Figure 5.2).

Answer Sheet

Lab 5: Microscopes and Cells

Name: _____   Section: _____

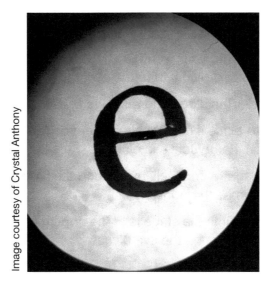

FIGURE 5.2 Letter 'e' Slide.

7. Using the fine focus knob bring the "e" into sharper focus. Draw what you see at the 4× objective in the labeled drawing circle at the back of the lab.

8. Take note of the orientation of the letter "e" when you look into the microscope. Is it in the same orientation as when you look directly at the slide?

> Yes or No

9. Rotate the nosepiece so that the 10× objective is now in place.

10. **Only a fine adjustment with the fine focus knob should be necessary**. If the specimen is mostly in focus after changing magnification this means that the microscope is **parfocal**. Do not use the coarse focus knob to adjust the focus. Draw what you see in the labeled drawing circle at the back of the lab.

11. **Rotate the nosepiece so that the 40× objective is now in place**. Once again you should only have to use the fine adjustment knob to focus the image. Draw what you see in the labeled drawing circle at the back of the lab.

Part III – Working Distance

© Joey Chung/Shutterstock.com

1. Using the e slide measure the **working distance**, the distance between the slide and the bottom of the objective, using a plastic metric ruler. When the e is in focus what is the working distance in mm (# of small lines on ruler) at 40×, 100×, and 400× **total magnification**?

   40×_____ mm   100×_____ mm   400× (estimate)_____ mm

   Did the working distance increase or decrease as you increased magnification?

   [   ]

2. Remove the slide and return it to the correct tray.

Part IV – Ruler Slide (Field of View)

1. Place ruler slide on the stage.

2. To demonstrate the concept of **field of view**, observe how many millimeters of the ruler are in view at 40×, 100×, and 400× **total magnifications**. In the box below indicate if the length of the field of view changed with an increase in magnification?

   | 40× _____ mm   100×_____mm   400×_____mm |
   | Increase   or   Decrease (circle one) |

3. Remove the slide and return it to the correct tray.

Part V – Salt Crystal Procedure

1. Place a slide of salt crystals on the stage. Bring this slide into focus as you have done before. Using the 10× objective, open and close the iris diaphragm. At what point is the **contrast the greatest**?

   [   ]

2. Does contrast increase or decrease as the diaphragm is closed?

   [   ]

3. Draw the salt crystals, using the 10 objective (100× total magnification) with the diaphragm open and with the diaphragm closed on the labeled drawing sheets at the end of the lab.

4. Remove the slide and return it to the correct tray.

Part VII – Animal Cells

Animals are multicellular organisms that ingest organic matter. They are composed of cells that can be categorized into four major tissue groups; epithelia, connective, muscle, and nervous tissue. Today we will be examining epithelial cells that form the lining of your check cells (Figure 5.3).

To obtain a specimen, follow the procedure below:

a. With a clean toothpick, gently scrape the inside of your cheek several times

b. Roll the scraping into a drop of water on a clean microscope slide.

c. Add a drop of methylene blue, and cover with a coverslip. Discard the toothpick in the 10% bleach solution provided.

d. View your cells using the 40× objective lens.

e. Draw your check cells in the labeled circle on the drawing sheets at the end of the lab.
   Label the following structures:
   - The Cell Membrane is the boundary that separates the cell from its surroundings.
   - The Nucleus is the large, circular organelle near the middle of the cell.
   - The Cytoplasm is the granular contents of the cell, exclusive of the nucleus.

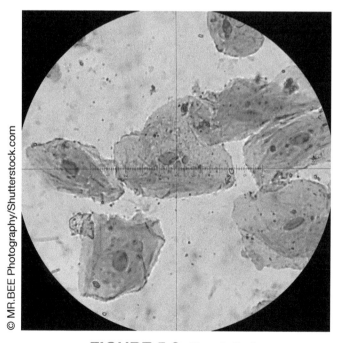

FIGURE 5.3 Cheek Cell.

Part VIII – Clean Up

- Replace the prepared slides in the proper slide box and dispose of the used glass slides and cover slips in the used microscope disposal can.

- Throw away your used lens paper.

- Return all other lab equipment and supplies to the lab cart.

- Wipe down your lab station.

- When you are done using the microscope turn it off, set the light to the lowest setting, remove the slide, rotate the 4× scanning objective into place, unplug the microscope, center the stage, and place the stage at the lowermost point. Wrap the cord with a rubber band. Get a signature from your instructor or lab aide. Next, place the microscope in the cabinet with the arm facing outwards. Make sure that you place it in the correct numbered cabinet. Be sure to carry it properly! Get signed off by your instructor or IA.

　　　　_____ Signature that the scope was put away correctly
　　　　_____ # of the scope that you used

Submit all of your drawings on the drawing circles sheets provided, properly labeled with the name of slide/specimen and the magnification you observed the slide at.

Microscope Drawings

1. Drawing of letter e, as seen through the 4× objective

　　　Total Magnification_____×

2. Drawing of letter e, as seen through the 10× objective

　　　Total Magnification_____×

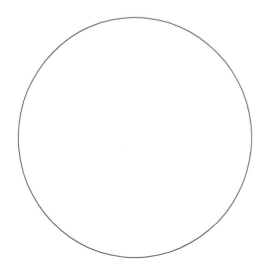
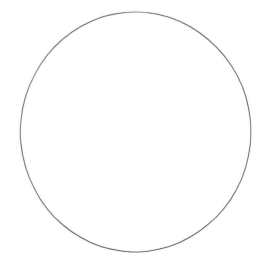

3. Drawing of letter e, as seen through the 40× objective

   Total Magnification_____×

4. Drawing of salt crystals, with diaphragm open, as seen through the 10× objective

   Total Magnification_____×

5. Drawing of salt crystals, with diaphragm closed, as seen through the 10× objective

   Total Magnification_____×

6. Drawing of the cheek cell, with Methylene Blue stain as seen through the 40× objective. **Label the cell membrane, cytoplasm, and nucleus**

7. **List 5 other organelles** in a Eukaryotic cell, that can't be seen in this view. **Describe the function of each of these organelles.**

Total Magnification_____×

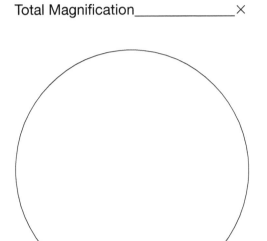

End of Lab Questions

1. You place a new slide on the stage of your microscope and look through the eyepieces. Unfortunately, even though the microscope is in focus you still don't see the specimen. **Describe, step by step, how you would use the stage control knob and focus knobs to bring the specimen into focus.**

2. You have a specimen in focus with the 4× objective and you want to change to the 40× objective to view the specimen at a higher magnification. **Explain, step by step, how you would switch the objective and keep the specimen in focus while increasing the magnification.** Keep it simple!

3. You are looking at a specimen but there is not enough contrast to see it clearly. **How would you go about increasing the contrast?** Again, keep it simple!

4. Define, in your own words, **contrast** and **differential staining**.

Laboratory 6

Diffusion and Osmosis

Objectives

Be able to do the following:

- Explain why a particular material diffuses in a particular direction
- Determine the net direction of diffusion
- Differentiate between diffusion and osmosis
- Describe the influence of temperature on the rate of osmosis
- Describe the influence that varying the concentration of solute and solvent has on the rate of osmosis.

Introduction

Although you may not know what diffusion is, you have experienced the process. Can you remember walking into the front door of your home and smelling a pleasant aroma coming from the kitchen? It was diffusion of molecules from the kitchen to the front door of the house that allowed you to detect the odors. Diffusion is defined as the net movement of molecules from an area of greater concentration to an area of lesser concentration until the concentration everywhere is equal. The movement in one direction minus the movement in the opposite direction determines the direction of net movement. To better understand how diffusion works, let's consider some information about molecular activity.

The molecules in a gas, a liquid, or a solid are in constant motion because of their kinetic energy. Moving molecules are constantly colliding with each other. These collisions cause the molecules to move randomly. The higher the concentration of molecules in one region, the greater the number of collisions. Some molecules are propelled into the less concentrated area and others are propelled into the more concentrated area. Over time, however, there will be more collisions in the highly concentrated area, resulting in more molecules being propelled into the less concentrated area. Thus, the net movement of molecules is always from more tightly packed areas to less tightly packed areas.

Diffusion occurs when there is a difference in concentration from one region to another or from one side of a membrane to another. A difference in the concentration of molecules over a distance is called a concentration gradient. When the molecules become uniformly distributed, they have reached dynamic equilibrium, in which the number of molecules moving in one direction is balanced by the number moving in the opposite direction. It is dynamic because molecules

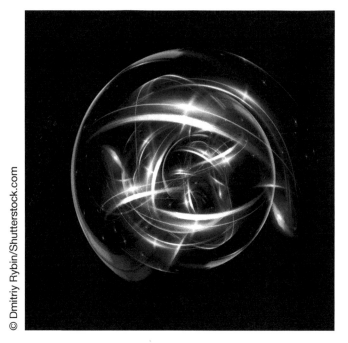

FIGURE 6.1 Molecular motion.

continue to move, but because motion is equal in all directions and there is no net change in concentration over time, equilibrium exists. The process of diffusion occurs in both living and nonliving systems. Biologically speaking, diffusion is responsible for the movement of a large number of substances, such as gasses and small uncharged molecules, into and out of living cells. The direction of diffusion is always from where there were originally more molecules to where there are fewer. This is similar to the scattering of a crowd of people leaving a theater. Many of the individuals move from the theater to the outside, but some go back to retrieve their coats or popcorn. The net movement, however, is the movement of the individuals leaving the theater minus the movement of those returning. Imagine that your instructor opens a bottle of ammonia in a corner of the room. The bottle would have the highest concentration of ammonia molecules in the room; the individual ammonia molecules would move from this area of highest concentration to where they are less concentrated (Figure 6.2).

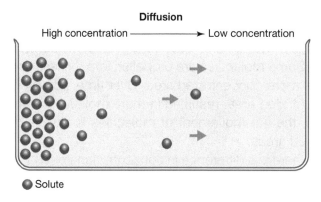

FIGURE 6.2 Solute transport is from the left to the right; movement of the solutes is due to the concentration gradient.

Although you could not actually see this happening, ammonia molecules would leave the bottle and move throughout the air in the room because of molecular movement. You could detect this by the odor of the ammonia. If you compare the relative number of ammonia molecules in the bottle to those dispersed in the room, you would be dealing with what is called relative concentration. Relative concentration compares the amount of a substance in two separate locations. Whenever there is a difference in concentrations of a substance, you can predict the direction that most of the molecules will move.

You can predict that when the bottle is first opened, ammonia molecules will move from the area of higher concentration (the bottle) to the region of lower concentration (the air in the room). Soon, however, the molecules of ammonia will mix with the air molecules in the room. Because the ammonia molecules are moving randomly, some of them will move from the air back into the bottle.

As long as there is a higher concentration of ammonia molecules in the bottle, more of them move out of the bottle than move in. One way of dealing with the direction of movement is to compare the number of molecules leaving the bottle with the number reentering the bottle. This is called the net amount of movement. The movement in one direction minus the movement in the opposite direction is the direction of net movement. If, for example, 100 molecules of ammonia leave the bottle and 10 reenter during that time, the net movement is 90 molecules leaving the bottle. Ultimately, the number of ammonia molecules moving out of the bottle will equal the number of ammonia molecules moving into it. When this point is reached, the ammonia molecules are said to have reached dynamic equilibrium.

When several kinds of molecules are present, consider only one case of diffusion at a time even though several different types of molecules are moving. For example, consider the exchange of gasses between the lungs and blood. In the lungs, there are a series of tubes that transport gases. These tubes divide into smaller and smaller branches and eventually end at a series of small alveolar sacs. Adjacent to these sacs are a number of capillaries containing blood. By the process of diffusion, there is an exchange of oxygen and carbon dioxide between the alveolar sacs and the blood in the capillaries (Figure 6.3). Carbon dioxide (CO_2)

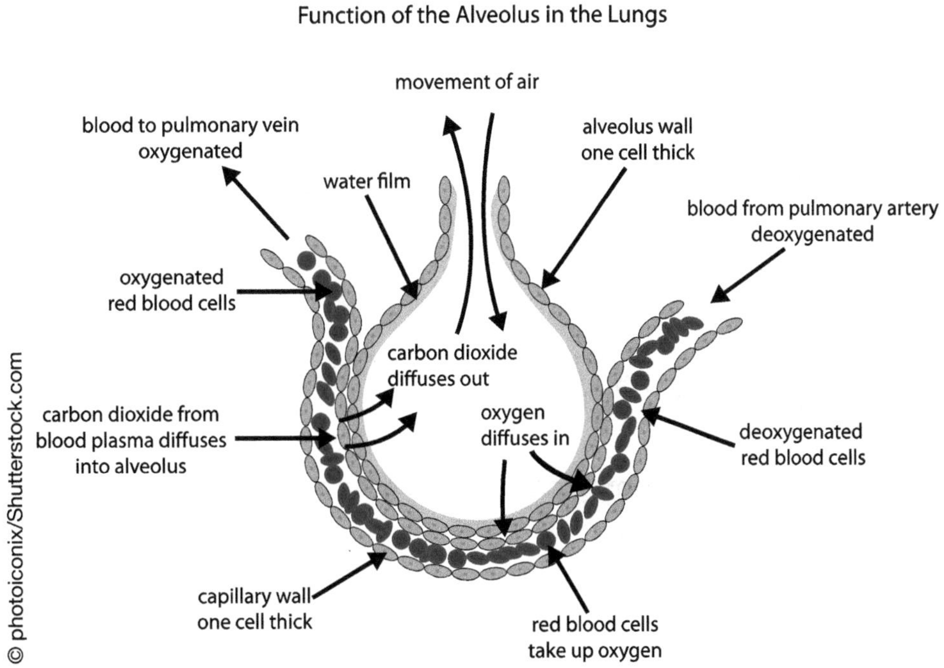

FIGURE 6.3 Diffusion of oxygen and carbon dioxide.

will follow its concentration gradient into the alveolus; oxygen (O_2) will follow its concentration gradient into the capillary.

Another example of diffusion is sucrose dissolving in water. When sucrose molecules and water mix, a solution is created. A solution is any mixture where two or more different types of molecules are evenly dispersed throughout the system. The dissolved substance called the solute characterizes a solution. Sucrose in this example is the solute. The substance in which the solute is dissolved is called the solvent. In this example (and in biological systems), water is the solvent.

The diffusion of water across a semi-permeable membrane is called osmosis. A semi-permeable membrane allows certain molecules to cross, but prevents others from crossing and can be a natural cell membrane or a thin sheet of material such as dialysis tubing. Some membranes are permeable only to water which means water molecules may freely diffuse across the membrane, but other types of molecules may not. Osmosis will occur when solute concentration on one side of the semi-permeable membrane is higher than the other side. Net movement of water will be toward the side of the membrane with a higher concentration of solute as shown in Figure 6.4.

In each of the previous examples, the net movement was a result of diffusion of molecules from a place of higher concentration to a place of lower concentration. The speed at which diffusion occurs is related to the temperature of the molecules and the concentration difference between the areas of high and low concentration.

The kinetic molecular theory states that all substances are made up of molecules that occupy space and are constantly in motion. This exercise helps you examine some phenomena related to this motion of molecules.

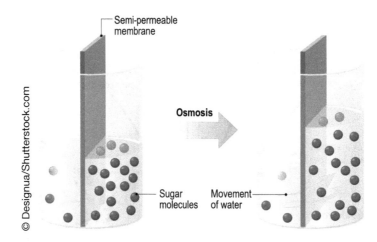

FIGURE 6.4 Movement across a semi-permeable membrane.

Lab 6: Pre-Lab Questions

Name: _____

Read through **all** of the lab handout and answers these questions **before** coming to class.

1. _____ is defined as the net movement of molecules from an area of greater concentration to an area of lesser concentration until the concentration is equal everywhere.
   a. Diffusion
   b. Osmosis
   c. Semi-permeable membrane
   d. Concentration gradient

2. A _____ is the difference in the concentration of molecules over a distance.
   a. Diffusion
   b. Osmosis
   c. Semi-permeable membrane
   d. Concentration gradient

3. A _____ is a thin sheet of material that selectively allows certain molecules to cross.
   a. Diffusion
   b. Osmosis
   c. Semi-permeable membrane
   d. Concentration gradient

4. Which dyes will we be using in the First diffusion experiment?
   a. Bromothymol Blue
   b. Xylene Cyanol
   c. Malachite green
   d. Orange G

5. The movement of water across a semi-permeable membrane is called?
   a. Diffusion
   b. Osmosis
   c. Kinetic movement
   d. Concentration gradient

6. What type of tubing will we be using as a semi-permeable membrane?
   a. Dialysis tubing
   b. Aquarium tubing
   c. Scientific tubing
   d. Osmosis tubing

7. Which of the following is **not** a starch concentration that we will be testing for an effect on diffusion?
   a. 25%
   b. 50%
   c. 100%
   d. We will be testing all of these concentrations

8. Which carbohydrate will we be filling the dialysis tubing with?
   a. Glucose
   b. Starch
   c. Sucrose
   d. Glycogen

9. How often will you measure the diameter of your growing dye circle?
   a. Every minute
   b. Every 5 minutes
   c. Every 10 minutes
   d. Every 20 minutes

10. During the experiment with dialysis Tubing, what will you be measuring and recording?
    a. Time of the reaction
    b. Length of the dialysis bag
    c. Weight of the dialysis bag
    d. Density of the dialysis bag

Procedure

Experiment #1: Diffusion in a Semi-Solid Matrix

In this exercise you will examine the movement of dyes diffusing within a semi-solid medium called agar, which is very similar to the gelatin that you eat. The agar forms a gel-like matrix when mixed with water, and is clear so you can see into it. Molecules diffuse through the water-filled channels in the agar matrix. You will be given 2 different dye molecules, along with their molecular weights, and will be asked to determine the relationship between size and the rate of diffusion.

☠ These molecules are toxic! Use gloves!

1. Before you begin develop a hypothesis about the relationship between dye size and diffusion speed in the results section.

2. Put a piece of paper over the space you will be working in to catch any spills. A paper towel works well too.

3. Place your Petri dish with agar on the piece of white paper.

4. Place a flat metric ruler between the Petri dish and the piece of white paper to measure diffusion in action.

5. Place 15 drops of Malachite green from the dropper bottle into ONE of the holes in your petri dishes agar (see image below).

6. In the other hole add 15 drops of Potassium Permanganate. Make sure to keep track of which hole has which dye in it. If it looks like your wells can hold additional drops, add them. However, add an equal amount of drops to each well.

7. After you have added the dyes try not to move the plate because the dyes could spill out of the holes.

8. Measure the initial width of each colored circle, in millimeters (mm) NOT centimeters (cm).

9. At 10-minute intervals measure the width of each expanded colored circle in mm for a total of 40 minutes.

10. Record your results in the Tables 1 and 2 in the results section.

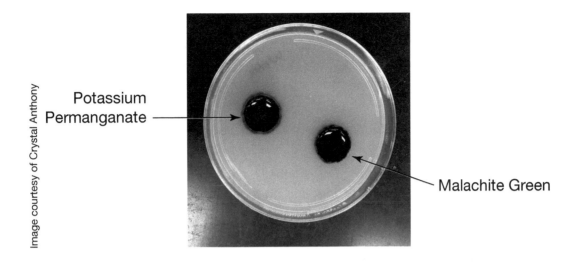

Experiment #2: Osmosis, Diffusion and the Effect of Concentration

In this exercise you will measure osmosis and the diffusion of small molecules through dialysis tubing, an example of a semi-permeable membrane. Dialysis tubing is used to separate substances in a solution due to their difference in molecular weight (size). The size of the minute pores (or holes) in the tubing determines which substances can pass through the membrane. The size of the pores of our dialysis tubing is 12,000 to 14,000 Daltons (very tiny!). Starch solutions or tap water will be placed inside your dialysis tubing. This tubing will be placed in a beaker containing water and I_2KI (iodine). At 5-minute intervals, you will weigh the bag and examine the solution to determine which molecules passed through the semi-permeable membrane.

Lugol's iodine (I_2KI) is used to test for the presence of starch. When it is added to an unknown solution, the solution turns purple if starch is present. If no starch is present, the solution remains a pale yellow-amber color. We have already used Lugol's iodine to reveal the starch grains in potato cells.

Prelab observation—At the front of the lab is an example of the reaction between Lugol's iodine and starch. Before you begin the experiment familiarize yourself with the color change

1. Before you begin record your group's hypothesis about the relationship between concentration and osmosis in the results section.

2. Put a piece of paper towel over the space you will be working to catch any spills.

3. Obtain three pieces of dialysis tubing and soak them in tap water for about 1 minute (this may have been done for you already).

4. Form each of the pieces of tubing into a tubular sac by rubbing them back and forth between your fingers. Shake off the excess water, fold over one end of the tubing and tie it securely (two or more knots) with a piece of dental floss.

5. Shake the squirt bottle of 100% starch to mix the contents. Fill the first tubular sac with 15 mL of the 100% starch solution. Use the provided graduated cylinder and funnel to get an accurate amount without spilling too much on the desk. Tie the top of the sac making sure to leave room for diffusion and osmosis to occur. Rinse the outside of the sac to remove any starch and pat it dry.

6. Rinse out the graduated cylinder and funnel with tap water.

7. Shake the squirt bottle of 50% starch to mix the contents. Fill the second tubular sac with 15 mL of the 50% starch solution. Use the provided graduated cylinder and funnel to get an accurate amount without spilling too much on the desk. Tie the top of the sac leaving space, rinse, and pat it dry.

8. Rinse out the graduated cylinder and funnel with tap water to get rid of any leftover starch before continuing.

9. Fill the third tubular sac with 15 mL of tap water. Use the provided graduated cylinder and funnel to get an accurate amount without spilling too much on the desk. Tie the top of the sac leaving space, rinse, and pat it dry.

10. Using the ends of the dental floss that you tied the bag closed with, tie each end of your prepared dialysis tubes to a glass stir rod. Your bags should be hanging down off of the glass rods like pigs about to be roasted.

11. Using an electronic balance determine the initial weight of each tube with the glass rod attached. Record your weights in Table 4 in the results section. Be sure to zero the balance with the weigh boat before each measurement. Ask for assistance with the balance if it doesn't seem to be working right.

12. Fill three beakers almost full with tap water and ask your instructor or instructional aide to add the appropriate amount of Lugol's iodine (I_2KI) to the tap water.

13. Record your initial observation of the color of the starch solution inside the dialysis tube and the color of the water in the beaker in Table 3.

14. Place each glass rod on top of a beaker filled with tap water and Lugol's iodine so that the dialysis bag is submerged in the liquid. Be sure to keep track of which bag is in which beaker.

15. After 5 minutes, individually remove each sac with the glass rod attached and gently squeeze to assess any changes in firmness. Observe the size, shape, and firmness of each sac.

16. Weigh each bag with the glass rod attached using the electronic balance used in step 10. Record your weights in Table 4. Also record your color observations in Table 3.

17. During this experiment you can tell whether water is entering or leaving the dialysis bag by the amount of weight that is gained or lost.

18. Return each sac to its appropriate beaker of water for another 5-minute interval. Repeat your observations and measurements every 5 minutes for 20 minutes (measurements should be taken at 0, 5, 10, 15, and 20 minutes).

19. After taking all of your measurements you will subtract the initial weight at time 0 from the weights at the other time points.

20. Graphs the adjusted weights to determine which concentration of starch in the dialysis bags resulted in the most osmosis.

Clean Up

1. Throw away dialysis bags and dye dishes into trash can
2. Place iodine water in correct disposal container
3. Rinse out glassware and glass rods
4. Bleach your entire lab station, including sink
5. Return lab supplies to designated area

Answer Sheet

Lab 6: Diffusion

Name: _____ Section: _____

1. Using the Reference Chart below, develop a hypothesis about the relationship between dye size and diffusion. Which dye will diffuse the fastest through the agar?

   Reference Chart

Dye	Size (g/mole)
Malachite Green	929
Potassium Permangante	158

   Your hypothesis:

TABLE 1 Molecule – Malachite Green

Time (minutes)	Width of Circle/Halo (mm)
Initial	
10	
20	
30	
40	

TABLE 2 Molecule – Potassium Permanganate

Time (minutes)	Width of Circle/Halo (mm)
Initial	
10	
20	
30	
40	

2. Based on your results in Tables 1 and 2 was your hypothesis about the relationship between size and speed of diffusion supported or rejected? What does this tell you about the relationship between size and diffusion?

3. Develop a hypothesis about the relationship between concentration and osmosis. Which concentration of starch in the dialysis bag (0%, 50%, or 100%) will cause the most osmosis to occur (most weight gained)?

Your hypothesis:

TABLE 3 Effect of Concentration on the Rate of Osmosis

Starch Concentration	Initial Weight (in grams)	Weight at 5 minutes	Weight at 10 minutes	Weight at 15 minutes	Weight at 20 minutes
0% (Tap water)					
Weight of bag minus the initial weight →	0				
Color of the solution inside the dialysis bag					
Color of the solution in the beaker					
50% Starch					
Weight of bag minus the initial weight →	0				
Color of the solution inside the dialysis bag					
Color of the solution in the beaker					
100% Starch					
Weight of bag minus the initial weight →	0				
Color of the solution inside the dialysis bag					
Color of the solution in the beaker					

- Construct a line graph below using the weights of the bags minus the initial weight in Table 3. Use three different colors of pencil, one for each concentration. All of the lines should start at 0 grams
- **Make sure to label the x-axis and y-axis, include a legend, and title your graph.**

4. Based on your graph above was your hypothesis about the relationship between concentration and osmosis supported or rejected? What does this tell you about the relationship between concentration and osmosis?

5. The experiment you set up also tested whether three different molecules (water, I_2KI - Lugol's iodine, and starch) can diffuse across the semi-permeable dialysis tubing. Below is a picture of your setup. For the 100% starch situation label the picture with the initial locations of the I_2KI, starch, and tap water? Draw arrows to identify which substances moved during the experiment in which direction.

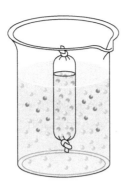

Reference Chart of Molecule Sizes

Item	Size in Daltons
Dialysis Tubing Pore Size	12,000–14,000
Lugol's Iodine (I_2KI)	121
Starch	20,000–100,000
Water	10

6. Review the reference chart above. Compare the size in Daltons of the tubing pores (small holes) and Lugol's iodine. Can Lugol's iodine pass through the pores in the dialysis tubing? Why or why not?

7. Review your observations in Table 3. During the experiment did the color of the liquid inside the dialysis bag change? Does this demonstrate that the Lugol's iodine can pass through the semi-permeable membrane of your dialysis tube? Why or why not?

8. Review the reference chart above. Compare the size in Daltons of the tubing pores (small holes) and Starch. Can starch pass through the pores in the dialysis tubing? Why or why not?

9. Review your observations in Table 3. Based on the color changes you observed did the starch stay in the dialysis bag or did it diffuse out? Explain.

10. Review the reference chart above. Compare the size in Daltons of the tubing pores (small holes) and water. Can water pass through the pores in the dialysis tubing? Why or why not?

11. In this example there was 100% starch in the dialysis bag and 0% starch in the beaker solution. Is the beaker solution isotonic, hypertonic, or hypotonic? Based on the type of solution you choose should the water flow into the bag or out of the bag? Explain.

Laboratory 7

Tissue Organization: Histology

Objectives

Be able to do the following

- Explain what the different tissue types are
- Describe what features makes one tissue unique from another
- State locations and functions of various tissue types
- Distinguish the various tissue types

Introduction

Histology is the study of tissues. Tissues are a groups of cells and material around them that work together to perform a particular function. There are 4 main types:

1. Epithelial: Covers body surfaces; lines organs, body cavities, and ducts. It also forms glands. Examples: Epidermis and Lining of Digestive Tract
2. Connective: Connects and supports the body. Examples: Ligaments and Tendons
3. Muscle: Generates force for movement. Examples: Skeletal, Cardiac, and Smooth
4. Nervous: Detects changes and responds with nerve impulses; helps maintain homeostasis. Examples: Brain, Spinal Cord, and Nerves

Procedure

Choose **3 Epithelial tissues and 4 Connective tissues (Loose/Dense and from cartilage, blood and bone), 2 Muscle Tissues, and a Neuron**. Find the appropriate slide, view it under the microscope, and then call your **Instructor or IA over to verify and initial that you found at least 2**. You will then draw it. **Be sure to label at least 2 features of the cell or tissue**. Be sure to ask for help if you are unsure if you are looking at the correct thing, many different tissues can be found on one slide even though it's labeled as one type. **Note: You are still responsible for identifying all the tissues and their location and function for the practical exam.**

A. Epithelial Tissue (Epithelium)–Involved in protection, filtration, secretion, absorption, and excretion. Cells are arranged in continuous layers. Avascular (no blood supply). Nutrients come from diffusion through the connective tissue. Has a nerve supply. Is constantly renewing and repairing.

Shapes:

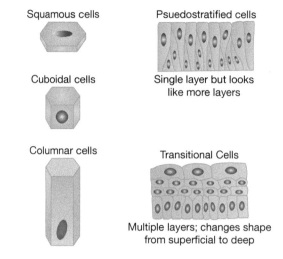

Layering:

Simple Epithelium: One layer Functions: *diffusion, *osmosis, *filtration, *secretion, *absorption

Stratified Epithelium: Two or more layers Functions: *Protection, *area with lots of wear and tear (more durable) *deeper cells undergo cell division, *cells at surface are dead (ex: skin)

Types:

1. Simple Squamous - Thin Membrane; Lung (respiratory), Kidney

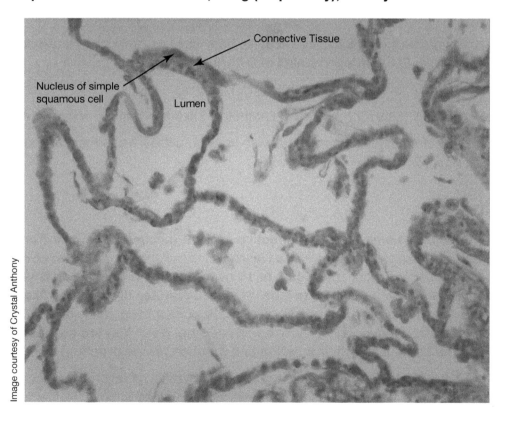

2. Simple Cuboidal - Thyroid gland, Kidney

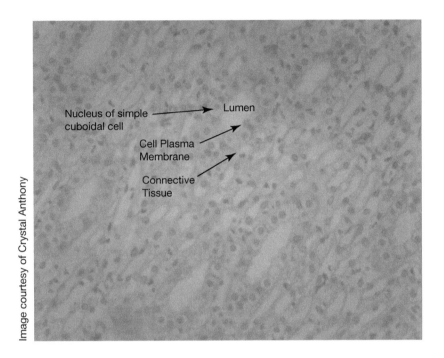

3. Simple Columnar – Most Common; Digestive, Respiratory, Thyroid Gland, Kidney

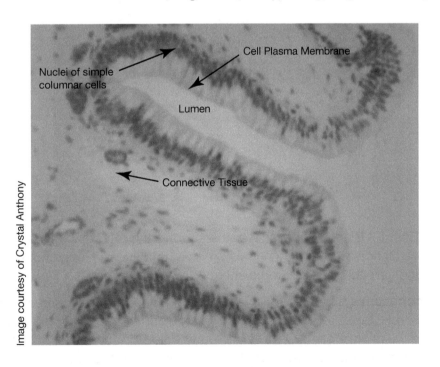

4. Stratified Squamous – Vagina, Skin, Tough, Wear And Tear

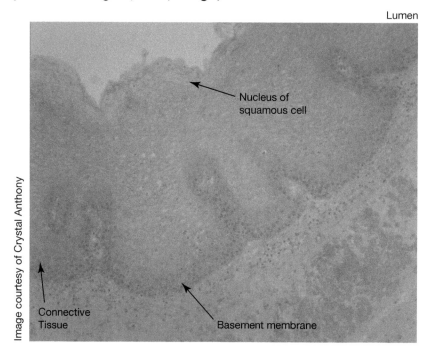

5. Psuedostratified Ciliated Columnar – Upper Respiratory Tract

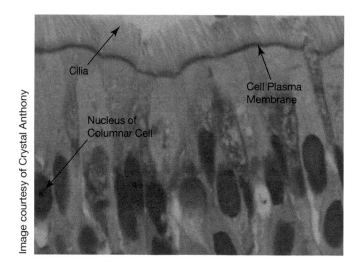

B. Connective Tissue – Binds structures together. Supports/protects/insulates organs. Involved in movement by attachment. Strengthens/reinforces other tissues. Separates structures. Transports nutrients and cells for growth. Stores energy (via adipose tissue).

Cells found in connective tissues: Fibroblasts make connective tissue fibers and ground substance of the matrix (Look like squamous cells); **Neutrophils and Lymphocytes** present with inflammation and infection; Plasma Cells secrete antibodies (proteins that neutralize foreign substances) {part of immune system (present with inflammation)}; **Mast Cells** are present near blood cells, produce histamine (chemical that dilates blood vessels in reaction to injury or infection) and heparin (inhibits blood clotting); **Adipocytes** are fat cells, they store triglycerides (fat); **Eosinphils** are white blood cells present with allergic reactions and parasitic infection; **Macrophages** are cells that destroy bacteria and debris via phagocytosis.

Fibers found in Connective Tissue: Collagen Fibers – Very strong yet flexible, resist stretching, are parallel bundles of a protein called Collagen; **Reticular Fibers** are thin collagen fibers coated with glycoprotein that provide strength; found as supporting framework of organs.

Types:

1. **Loose Connective Tissues** – Loosely woven fibers with many cells

    a. **Areolar Connective Tissue** - Forms fascia; attaches skin to tissues and organs; Many blood vessels and cells.

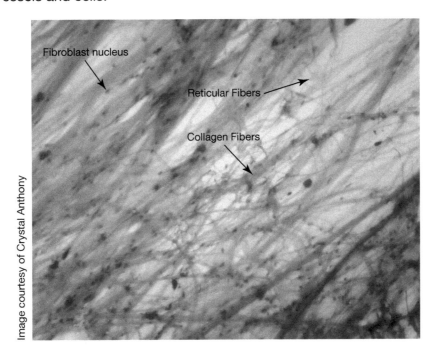

    b. **Adipose Tissue** - Stores fat/triglycerides; Provides energy and insulation.

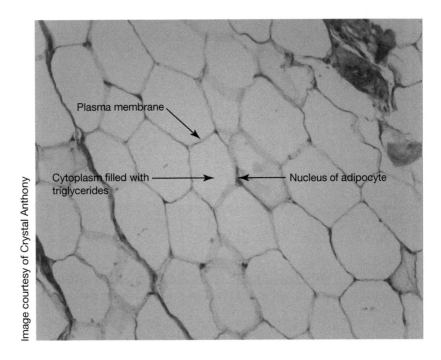

2. **Dense Connective Tissues** – Numerous thick fibers with few cells

   a. **Dense Regular (Fibrous) Connective Tissue** - Tendons and ligaments; Parallel fibers; Fibroblasts; Few RBC's (slow to heal)

   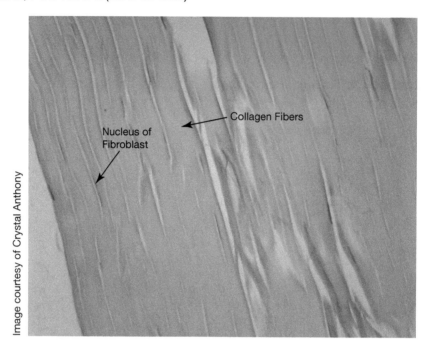

   b. **Dense Irregular Connective Tissue** - Dermis of skin; heart valves; Periosteum covering of bone; Protective capsule around organs

   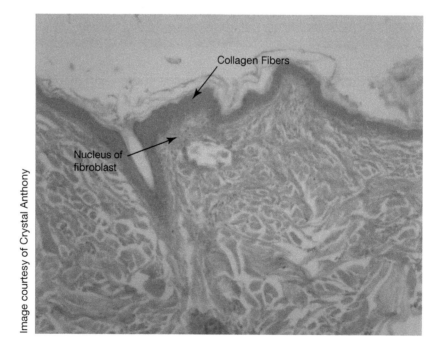

3. **Cartilage** is a connective tissue type that is embedded with a chemical that makes it stronger and thicker allowing it to withhold a lot of stress. Living cells with in the hyaline cartilage are called chondrocytes.

   a. Hyaline – most common, weakest; joints and growth plates

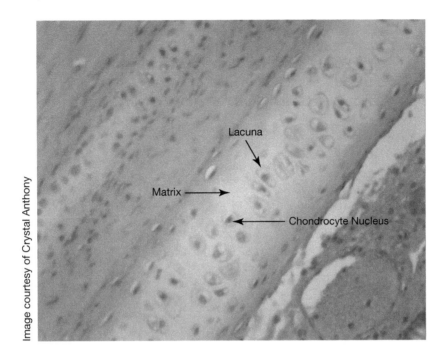

4. **Osseous Tissue (Bone)** – There are two types: Compact and Spongy bone, we will only look at compact histologically. **Compact bone** (long bones) have a circular structure called the Osteon. **Spongy bone** is found at the ends of bone and small bones. It is made up of columns of bone called trabeculae.

   a. Compact Bone – Main parts: Lamellae: concentric rings, which gives bone it's hardness; Lacunae: spaces between lamelle that house Osteocytes (living cells of bone); Canaliculi: canals for nutrients and wastes to travel through; Central Canal: contains blood vessels and nerves

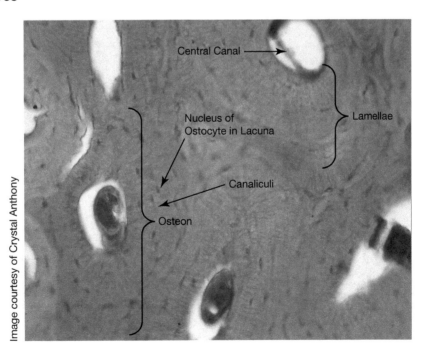

5. **Blood** – Connective tissue with a liquid matrix called plasma (dissolved nutrients, wastes, enzymes, hormones, respiratory gases and ions).
   Cells found in blood: Red blood Cells (Erythrocytes) – transport oxygen to body cells and removes CO_2; **White Blood Cells (Leukocytes)** – part of immune system involved in phagocytosis, and allergic reactions; **Platelets (Thrombocytes)** are involved in blood clotting.

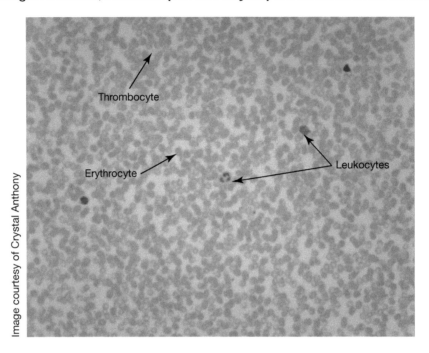

6. **Muscle tissue** – Cells that work together to generate force. Three types: **Skeletal, Smooth, and Cardiac**.
   a. **Skeletal Muscle Tissue**: Attached to bones; striated; under voluntary control; parallel fibers; many nuclei at periphery of cell.

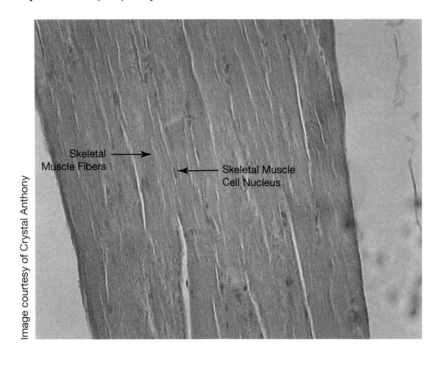

b. **Cardiac Muscle Tissue**: Wall of heart; striated; involuntary control; one to two nuclei; cells are attached end to end by intercalated discs.

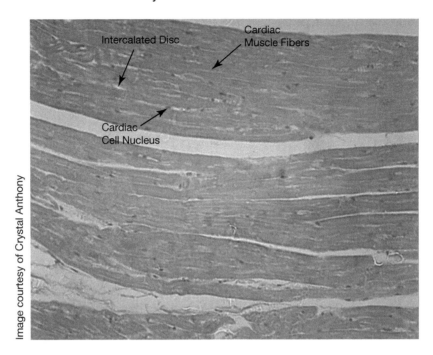

c. **Smooth Muscle Tissue:** In internal structures and organs; nonstriated; involuntary; spindle shaped with one centrally located nucleus.

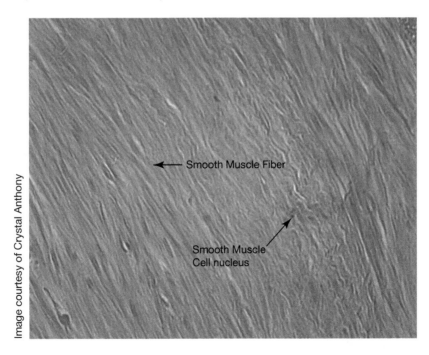

7. **Nervous Tissue- Neurons (Nerve Cells):** Convert stimuli to nerve impulses; **Neuroglia:** Do not generate nerve impulses, they support, protect, and provide nutrients to neurons.

   Neuron is made up of: Cell body (Soma) – contains nucleus and organelles; Dendrites – branches from the cell body; Axon – thin long tube that transports substances and the nerve impulse.

   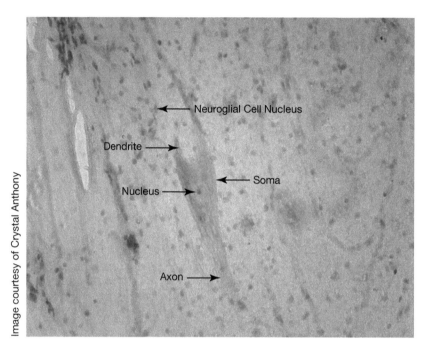

Lab 7: Pre-Lab Questions

Name: _____

Read through all of the lab handout and answers these questions **before** coming to class.

1. _____ is a tissue that is primarily covering and lining.
   a. Epithelial
   b. Connective
   c. Muscle
   d. Nervous

2. A _____ is a tissue that detects changes.
   a. Epithelial
   b. Connective
   c. Muscle
   d. Nervous

3. A _____ is a tissue that connects and supports the body.
   a. Epithelial
   b. Connective
   c. Muscle
   d. Nervous

4. A _____ is a tissue that generates force for movement.
   a. Epithelial
   b. Connective
   c. Muscle
   d. Nervous

5. Simple epithelium has which of the following function(s)?
   a. Diffusion
   b. Osmosis
   c. Protection
   d. All of the above

6. In this lab, how many tissue identifications will need to be signed off?
   a. 1
   b. 2
   c. 4
   d. 6

7. In this lab, how many features will you need to label on each of your drawings?
   a. 1
   b. 2
   c. 4
   d. 6

8. Stratified epithelium has which of the following function(s)?
   a. Diffusion
   b. Osmosis
   c. Protection
   d. All of the above

9. What is the main cell type found in loose connective tissue?
   a. Erythrocyte
   b. Myocyte
   c. Fibroblast
   d. Chondrocyte

10. What is the main cell type found in hyaline cartilage?
    a. Erythrocyte
    b. Myocyte
    c. Fibroblast
    d. Chondrocyte

Answer Sheet/Drawings

Lab 7: Tissue Organization: Histology

Name: _____ Section: _____

Choose 3 Epithelial tissues and 4 Connective tissues (Loose/Dense and from cartilage, blood and bone), 2 Muscle Tissues, and a Neuron. Find the appropriate slide, view it under the microscope, and then call your Instructor or IA over to verify and initial that you found at least 2. You will then draw it. Be sure to label at least 2 features of the cell or tissue. Note: You are still responsible for identifying all the tissues and their location and function for the practical exam.

1. Tissue: <u>Epithelial</u> Specific Type: _____
Location it can be found: _____
One Function of this tissue: _____
Magnification: _____ Signature _____

2. Tissue: <u>Epithelial</u> Specific Type: _____
Location it can be found: _____
One Function of this tissue: _____
Magnification: _____ Signature _____

3. Tissue: <u>Epithelial</u> Specific Type: _____
Location it can be found: _____
One Function of this tissue: _____
Magnification: _____ Signature _____

4. Tissue: <u>Connective</u> Specific Type: _____
Location it can be found: _____
One Function of this tissue: _____
Magnification: _____ Signature _____

5. Tissue: <u>Connective</u> Specific Type: _____
Location it can be found: _____
One Function of this tissue: _____
Magnification: _____ Signature _____

6. Tissue: <u>Connective</u> Specific Type: _____
Location it can be found: _____
One Function of this tissue: _____
Magnification: _____ Signature _____

7. Tissue: <u>Connective</u> Specific Type: _____
Location it can be found: _____
One Function of this tissue: _____
Magnification: _____ Signature _____

8. Tissue: <u>Muscle</u> Specific Type: _____
Location it can be found: _____
One Function of this tissue: _____
Magnification: _____ Signature _____

9. Tissue: <u>Muscle</u> Specific Type: _____
Location it can be found: _____
One Function of this tissue: _____
Magnification: _____ Signature _____

10. Tissue: Neuron Specific Type: _____
Location it can be found: _____
One Function of this tissue: _____
Magnification: _____ Signature _____

Laboratory 8

The Heart

Objectives

- Identify and locate the basic anatomy of the heart   ***Quiz at end of class***
- Demonstrate the flow of blood through the heart
- Demonstrate the flow of blood to/from the systemic and pulmonary circulation
- Correctly describe the cardiac system
- Describe and explain the physiology of events in the heart during the P wave, QRS complex, and T wave

Introduction

Read through the lab materials and answer the questions at the end of the lab. Be complete and thorough in your answers for full credit. Use the models and computers as resources.

You will be locating and identifying the following structures:

Right and Left Atria (Externally and Internally)

Right and Left Ventricles (Externally and Internally)

Coronary Vessels

Pulmonary Trunk

Right and Left Pulmonary Arteries

Right (Tricuspid) and Left (Bicuspid or Mitral) Interventricular Valves

Pulmonary and Aortic (Semilunar) Valves

Chordae Tendineae

Papillary Muscles

Brachiocephalic Trunk, Left Common Carotid, Left Subclavian Arteries

   (Note: Right Common Carotid and Right Subclavian Arteries branch off of Brachiocephalic Trunk)

Aorta – Ascending, Arch, Descending

Superior and Inferior Vena Cavae

Introduction

Cardiac Conduction System

- Myogenic – heartbeat originates within heart
- Auto rhythmic – depolarizes spontaneously regularly
- Conduction system
    - SA Node: pacemaker, initiates heartbeat, sets heart rate
    - Fibrous skeleton insulates atria from ventricles
    - AV Node: electrical gateway to ventricles
    - AV bundle: pathway for signals from AV node
    - Right and Left bundle branches: divisions of AV bundle that enter interventricular septum and descend to apex
    - Purkinje fibers: upward from apex, spread throughout ventricular myocardium

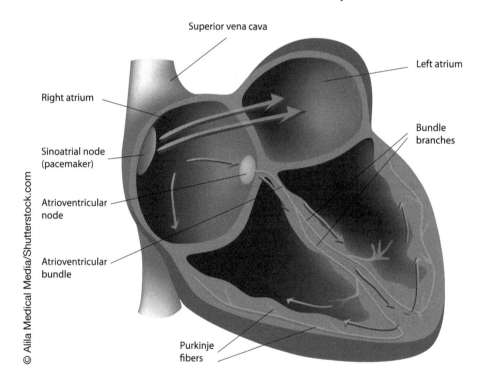

ECG

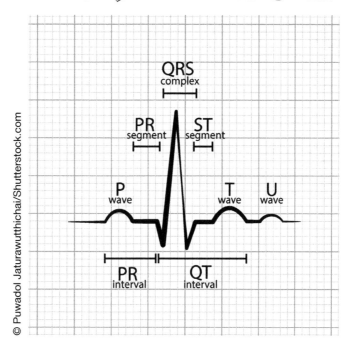

1. **Atria** begin to **depolarize**
2. **Atria depolarize;** atria contract (P wave)
3. **Ventricles** begin to **depolarize at apex; atria repolarize**
4. **Ventricles depolarize**; ventricles contract (QRS complex)
5. **Ventricles** begin to **repolarize** at apex (T wave)
6. **Ventricles repolarize**

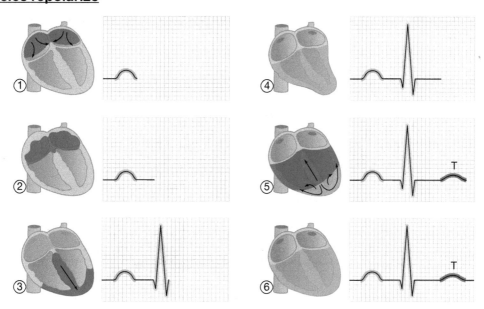

Cardiac Rhythm

- Systole = Ventricular contraction, Diastole = Ventricular relaxation
- Sinus Rhythm – set by SA node, adult @rest = 70 to 80 bpm
- Nodal Rhythm – set by AV node, 40 to 50 bpm
- Intrinsic Ventricular Rhythm 20 to 40 bpm
- Arrhythmia – abnormal cardiac rhythm
    - Heart block: failure of conduction system
    - Bundle branch block
    - Total heart block (damage to AV node)
- Tachycardia: persistent, resting adult HR >10
    - Stress, anxiety, drugs, heart disease or ↑ body temp.
- Bradycardia: persistent, resting adult HR <60
    - Common in sleep and endurance trained athletes

Contraction of Myocardium

- Myocytes (Cardiac cells) have stable resting potential of −90 mV
- Depolarization (very brief)
    - Stimulus open Na^+ gates, (Na^+ in) depolarizes to threshold, rapidly opens more Na^+ gates in a positive feedback cycle
    - Action potential peaks at +30 mV, Na^+ gates close quickly
    - Plateau: 200 to 250 ms, sustains contraction
    - Ca^{2+} channels in Sarcoplasmic Reticulum open and release Ca^{2+} into cytosol, which results in contraction
    - Repolarization: Ca^{2+} channels close, K^+ channels open, rapid K^+ out returns to resting potential

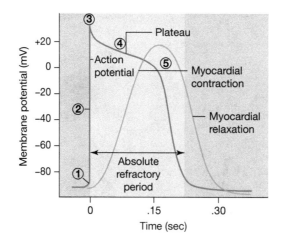

Laboratory 8: The Heart   93

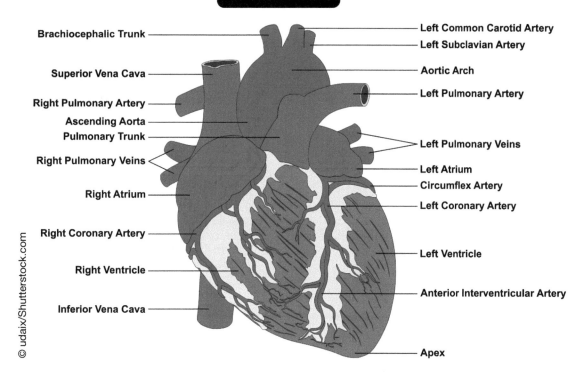

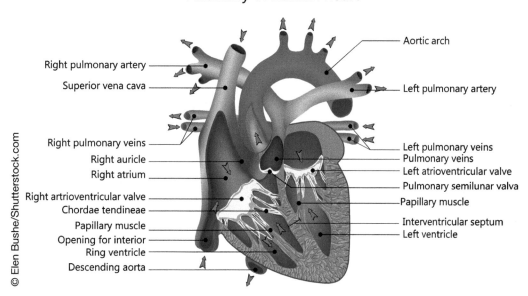

Anatomy of human heart

Lab 8: Pre-Lab Questions

Name: _____

Read through all of the lab handout and answers these questions **before** coming to class.

1. Blood flows from the left atrium into the _____.
   a. Right Ventricle
   b. Left Ventricle
   c. Pulmonary Artery
   d. Aortic Artery

2. Blood flows from the left ventricle into the _____.
   a. Right Atrium
   b. Left Atrium
   c. Pulmonary Artery
   d. Aortic Artery

3. Blood enters the heart from the body via the _____.
   a. Pulmonary Veins
   b. Pulmonary Arteries
   c. Aorta
   d. Inferior and Superior Vena Cavae

4. Cardiac tissue can be distinguished from Skeletal muscle tissue because it has.
   a. Striations
   b. No striations
   c. No nucleus
   d. Intercalated discs

5. The cardiac conduction system includes all of the following, EXCEPT?
   a. SA Node
   b. Purkinje fibers
   c. TR Node
   d. AV bundle

6. The ventricles contract during the?
   a. P wave
   b. QRS complex
   c. T wave
   d. QT segment

7. Systole is _____?
   a. A normal sinus rhythm
   b. An abnormal sinus rhythm
   c. Ventricular Contraction
   d. Ventricular Relaxation

8. Physiologically depolarization refers to?
   a. A movement of Na^+ ions across the myocyte cell membrane
   b. A movement of Ca^{++} ions across the myocyte cell membrane
   c. A movement of K^+ ions across the myocyte cell membrane
   d. None of the above

9. Which vessel does NOT branch off of the aorta?
   a. Right Subclavian Artery
   b. Left Subclavian Artery
   c. Left Common Carotid
   d. Brachiocephalic Trunk

10. Coronary Vessels supply blood to the?
    a. Somatic Cells
    b. Lungs
    c. Myocytes
    d. Chondrocytes

Answer Sheet

Lab 8: Heart

Name: _____ **Section:** _____

1. Using the model, locate the superior and inferior vena cavae. Using the model follow the path of blood flow through the heart. Then below, describe the flow of blood starting at the Superior and Inferior Vena Cavae and ending at the Superior and Inferior Vena Cavae. For example: The blood from the Superior and Inferior Vena Cavae dump into the Right Atrium, it then passes through the Right Atrioventricular (Tricuspid) Valve and into the…..

2. **Exterior Anatomy and Vessels.** Locate the following structures on the external surface of the heart model. Right Atrium, Left Atrium, Right Ventricle, Left Ventricle, Coronary Vessels, Pulmonary Trunk, Right and Left Pulmonary Arteries, Brachiocephalic Trunk, Left Common Carotid, Left Subclavian Arteries (Note: Right Common Carotid and Right Subclavian Arteries branch off of Brachiocephalic Trunk), Aorta – Ascending, Arch, Descending Superior and Inferior Vena Cavae

   The most superficial outside layer of the heart is the pericardium. The pericardium is a type of serous membrane secreting a slippery fluid around the outside of the heart. The pericardium has two layers called the parietal pericardium and the visceral pericardium. The parietal pericardium is a sac-like membrane enclosing the heart while the layer directly attached to the outside of the heart is the visceral pericardium.

3. Using the diagram below, place the correct letter for the various parts depicted in the diagram. Not all parts will be labeled.

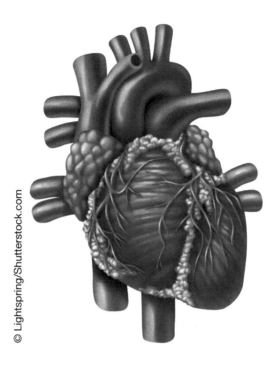

A. Arch of aorta
B. Pulmonary artery
C. Auricle
D. Pulmonary veins
E. Brachiocephalic artery
F. Superior vena cava
G. Coronary vessels
H. Inferior vena cava
I. Subclavian artery

4. **Interior Anatomy.** Locate the following structures on the internal surface of the heart model. Right Atrium, Left Atrium, Right Ventricle, Left Ventricle, Right (Tricuspid) and Left (Bicuspid or Mitral) Valves, Pulmonary and Aortic (Semilunar) Valves, Chordae Tendineae and Papillary Muscles

Which side has the greatest myocardium (heart muscle) thickness? _____

The very interior of the heart chambers is lined with a membrane called the endocardium. The extension of the endocardium forms the valves of the heart. In the right side of the heart between the atria and ventricle is the tricuspid valve while the bicuspid valve is between the atria and ventricle on the left side. Note that the valves have cord-like structures attached which are called chordae tendineae. These cords keep the valves in the proper orientation and keep blood from regurgitating back into the atria during ventricle contraction. The chordae tendineae attach to the endocardial layer by a cluster of muscles called the papillary muscles. These muscles contract during stretch and act to absorb some of the shock to the cords during ventricular contraction. Also within the endocardium are folds of heart tissue called trabeculae carnea. The trabeculae carnea is similar in function to the auricle in allowing for stretch and increasing surface area.

Do the trabeculae folds exist in the atria? _____

5. Describe the Cardiac conduction system. Where in the heart does the electrical current begin and where does it end? List the step by step flow anatomically.

6. Describe the events occurring in each of the following parts of the ECG reading: The P-wave, QRS complex, and T-wave.

7. Describe the physiological mechanisms (i.e., what cell processes/changes occur leading to an action potential) of the cell during the contraction of the myocardium.

Laboratory 9

Blood Cells

Objective

- Draw, describe and recognize the components of blood

Introduction

Granulocytes

Neutrophils: 60% to 70%; lightly noticeable granules; multi-lobed nucleus; bacterial infections- phagocytosis of bacteria and release antimicrobial chemicals

Eosinophils: 2% to 4%; large visible granules; bi-lobed nucleus; parasitic infections and allergies – phagocytosis of antigen–antibody complexes and release enzymes to destroy large parasites

Basophils: <1%; Many large violet granules; S-shaped nucleus (difficult to locate); Chickenpox, Allergies, Inflammation – secrete histamine (vasodilator): speeds flow of blood to an injured area and heparin (anticoagulant), promotes the mobility of other WBCs in the area

Agranulocytes

Lymphocytes: 25% to 33%; Large, round, dark violet nucleus; diverse infections and Immune responses - destroy cells, activate and coordinate other immune cells

Monocytes: 3% to 8%; largest WBC; kidney-shaped or horseshoe-shaped nucleus; viral infections and inflammation – phagocytize pathogens/debris and antigen-presenting cells (APCs)

Neutophil

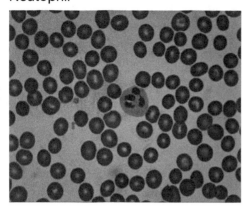

Eosinophil

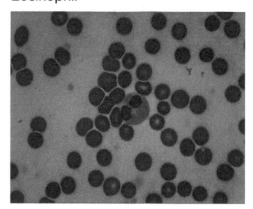

Basophil

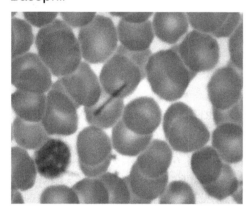

Lymphocyte

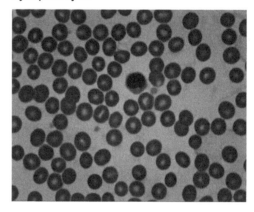

Monocyte

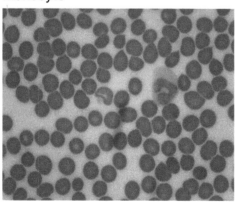

Procedure

1. Place the blood smear slide on the microscope. Draw what you see in the visual field. Therefore, the first drawing **must** have all basic components of blood; **Draw and label**: Erythrocyte, Leukocyte, and Thrombocyte. Be sure to provide a description.

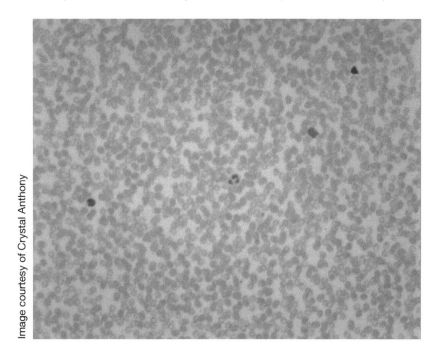

Image courtesy of Crystal Anthony

2. For the other three drawings you will focus in on a **specific Leukocyte** (of your choice). **Call the instructor or IA to verify** you found the Leukocyte you intend on drawing and have your paper signed. Draw the chosen Leukocyte and label it with its specific type (name). Also, describe its characteristics and list its primary function(s).

Lab 9: Pre-Lab Questions

Name: _____

Read through the entire lab an answer these questions before coming to class.

1. Neutrophils are involved in _____.
   a. Bacterial Infections
   b. Viral infections
   c. Parasitic infections
   d. Cellulose fermentation

2. Monocytes are involved in _____.
   a. Bacterial Infections
   b. Viral infections
   c. Parasitic infections
   d. Cellulose fermentation

3. Eosinophils are involved in _____.
   a. Bacterial Infections
   b. Viral infections
   c. Parasitic infections
   d. Cellulose fermentation

4. In a blood smear _____ are the most abundant Leukocytes.
   a. Monocytes
   b. Neutrophils
   c. Eosinophils
   d. Basophils

5. All of the following a Granulocytes EXCEPT _____?
   a. Monocytes
   b. Neutrophils
   c. Eosinophils
   d. Basophils

6. Agglutination will occur if you add _____ to a sample of A blood.
   a. Anti A serum
   b. Anti B serum
   c. Anti Rh serum
   d. Antigens

7. If the B antigen is present on the cell of a sample (i.e., The person has type A blood) then adding _____ will result in agglutination.
   a. Anti A serum
   b. Anti B serum
   c. Anti Rh serum
   d. Antigens

8. In this lab, you will be using _____ trays with 3 wells in them.
   a. 3
   b. 4
   c. 5
   d. 6

Answer Sheet

Lab 9: Blood Cells

Name: _____   **Section:** _____

○

Specific Type: _____Blood Smear_____
Magnification: _____ Signature _____
Description: _____
_____
_____

○

Specific Type of Leukocyte: _____
Magnification: _____ Signature _____
Description: _____
_____
_____

○

Specific Type of Leukocyte: _____
Magnification: _____ Signature _____
Description: _____
_____
_____

○

Specific Type of Leukocyte: _____
Magnification: _____ Signature _____
Description: _____
_____
_____

Laboratory 10

Blood Typing

Objective

- Understand and identify blood types

Introduction

There are 4 different types of human blood. There are determined by the presence or absence of proteins on the surface of the red blood cell (erythrocyte). These proteins vary in their chemical composition making them uniquely identifiable by your body's immune system. The primary proteins are called A and B antigens. About 40 years after A and B antigens had been discovered, another type of protein called the Rh factor was discovered. It is called Rh because it was initially discovered in Rhesus monkeys. Our final determination of blood type takes into account all of these proteins.

The chart below will help you to identify the blood types:

Protein(s) Present	Blood Type
A Antigen	A−
B Antigen	B−
A Antigen and B Antigen	AB
No Antigens	O−
A Antigen and Rh factor	A+
B Antigen and Rh factor	B+
A Antigen, B Antigen, and Rh factor	AB+
Rh factor	O+

As part of our immune system, to protect our bodies from foreign invaders, we have proteins that fight cells that do not belong to the organism. These proteins are called antibodies. When referring to the blood they are also called agglutinins because they cause agglutination, or clumping of the blood. This process allows us to test a sample of blood to determine its type. For example, a person with blood type A− will have antibodies for B antigens (Anti-B) and antibodies for Rh antigens (Anti-Rh). However, it must be noted that people with a negative blood type do not have Rh antibodies until after the first exposure to Rh+ blood (this can occur with a blood transfusion or when a Rh− mother gives birth to a Rh+ infant).

Using these principles, we can take a sample of blood and add different antibodies to it. If agglutination (clumping) occurs, then we know the blood sample contains that antigen on the surface of the cell. For example, if I add B antibodies (Anti-B serum) to a sample of blood and clumping occurs then I know the blood cells in the sample contain B antigens on their surface. If agglutination does not occur, then I know the sample does not have any B antigens on the surface of their cells. You will be using this technique and this logic to determine which blood was found at the scene of a crime.

Answer Sheet

Lab 10: Blood Type

Name: _____ **Section:** _____

Procedure You will be testing the following samples of blood:

1. Crime Scene
2. Victim
3. Suspect #1
4. Suspect #2
5. Suspect #3
6. Suspect #4

1. You will get 6 trays that have 3 wells in them. The wells are marked A, B, and Rh
2. Take a piece of paper towel (larger than the tray) and write Crime Scene on it.
3. Take the tray to the front of the class and place 3 to 4 drops of the blood sample into each well.
4. Add 3 to 4 drops of the A Anti-serum to the well labeled A.
5. Add 3 to 4 drops of the B Anti-serum to the well labeled B.
6. Add 3 to 4 drops of the Rh Anti-serum to the well labeled Rh.
7. Take a tooth pick or stir stick and stir each well (be sure to use a different stick for each well).
8. Determine if agglutination has occurred. If you are uncertain, move the written words under the tray, if it is difficult to read then agglutination has occurred.
9. Write the results in the Data Table. Determine the blood type.
10. Repeat with each of the samples.
11. When completed, rinse, wash, dry, and put away wells.

Data Table

*Write **YES** in the box if Agglutination occurred and **NO** if it did not.

Sample	Well A	Well B	Well Rh	Blood Type
Crime Scene				
Victim				
Suspect #1				
Suspect #2				
Suspect #3				
Suspect #4				

Answer the following questions:

1. Which sample's blood was found at the crime scene?

2. Does this mean the person whose sample was found at the crime scene is the killer?

3. What are some alternative explanations for the suspects blood being found at the crime scene?

4. Provide two additional types of evidence or information you would need in order to confidently conclude that the suspect whose blood was found at the crime scene is the killer.

5. Describe, in your own words, what are the features of blood cells that make one person's blood type different than another person's blood type? Why do you think humans evolved different blood types?

Laboratory 11

ELISA

Identify Patient Zero of a Zombie Apocalypse with the Power of an ELISA

Objectives

- Understand the basics of immunology, antibodies, and ELISA
- Explain Epidemiology and the study of the spread of a disease.
- Narrow down who were the initial carriers of the disease

Introduction

- Example disease in this lab: Zombiefication (antigen protein)
    - Method of infection: Bodily fluid contact, Airborne, Foodborne
    - Symptoms: Undead, Hungry for flesh
    - Treatment: None known
- Crash Course in Immunology and ELISA
    - Immune Response:

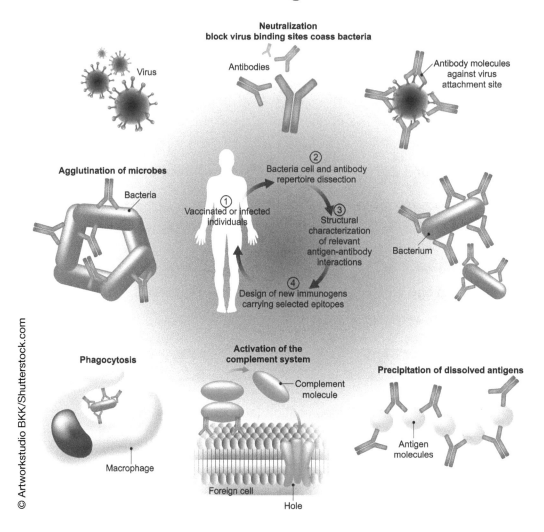

- ELISA tests are based on immune system antibody molecules (also a type of protein)

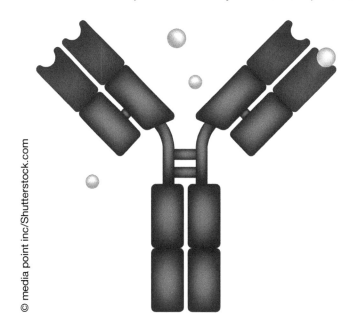

- **Overview of Procedure**

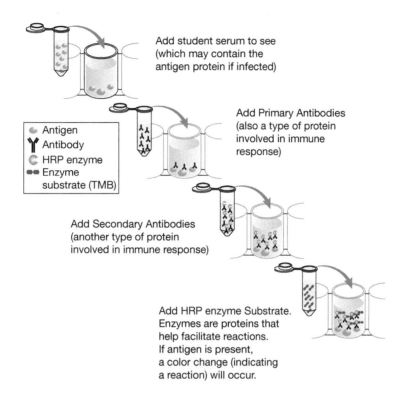

- **Procedure for Swapping Fluids:**

    1. **Label a yellow tube and plastic transfer tube with your initials**

    2. **Swap fluid with 1 person in your lab group (record who name and number)**

        a. **Add ALL your "body fluid" into the other student's tube of their "body fluid". They will gently mix the fluids together and give you back HALF the fluid (750 μL). Now you each have the same amount of fluid, which is a mix of both of your fluids.**

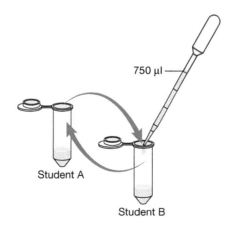

    3. **Swap fluid with a 2nd person, in another (closest) lab group (record who – name and number). Use the procedure in step 2a.**

4. **Swap fluid with a 3rd person, in another (farthest away) lab group (record who – name and number). Use the procedure in step 2a.**

5. **Discard the transfer pipet.**

Sharing Partner #1 _____

Sharing Partner #2 _____

Sharing Partner #3 _____

- **ELISA Time!**   **Testing for who is infected**

    1. Label 12 well strip. On first 3 put a "+" sign. On second 3 put a "−" sign. Next 3 with your Initials or ID# and Last 3 with your partner's initials or ID#.

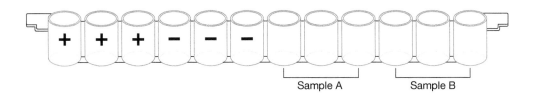

    2. **Use a fresh pipet tip to transfer 50 μL of the positive control (+) into the 3 "+" wells.**

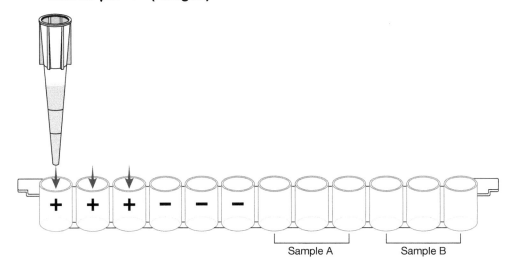

3. **Use a fresh pipet tip to transfer 50 μL of the negative control (−) into the 3 "−" wells.**

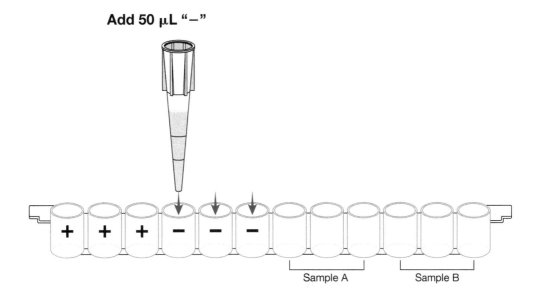

4. **Add 50 μL Bodily Fluid to each appropriately labeled well; use a fresh Pipet for each student's sample.**

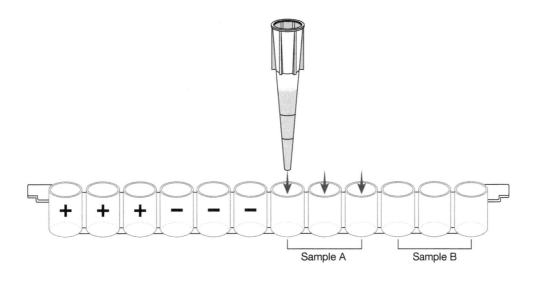

5. **Incubate 5 minutes – wait 5 minutes while all the proteins in the samples bind to the wells.**

6. **Tap out liquid onto paper towels. Gently tap the well strip a few times upside down. Make sure to avoid samples splashing back into wells. Discard top paper towels.**

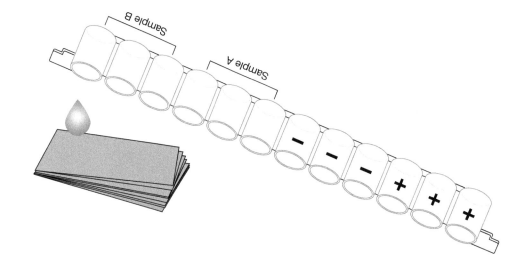

7. **Wash with buffer – use a fresh transfer pipet to fill each well with wash buffer, taking care not to spill over into neighboring wells. Note; use same transfer pipet for all washing steps.**

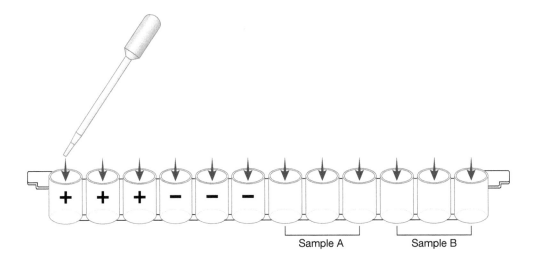

8. **Tip the well strip upside down onto the paper towels and tap. Discard top 2 to 3 paper towels.**

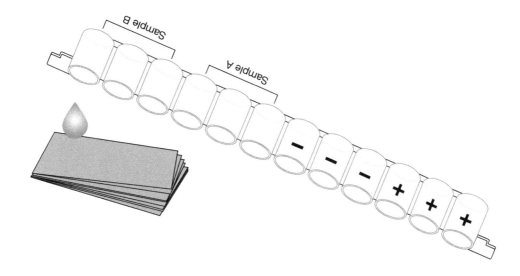

9. **Repeat wash steps 7 and 8.**

10. **Use a fresh pipet tip to Add 50 μL of "PA" (Primary Antibody) into all 12 wells of well strip.**

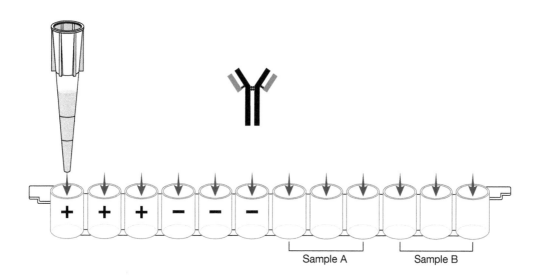

11. **Incubate 5 minutes – wait 5 minutes for the antibodies to bind to their targets.**

12. **Repeat wash steps 6, 7, and 8 – TWO times**

13. **Use a fresh pipet tip to Add 50 μL "SA" (Secondary Antibody) into all 12 wells.**

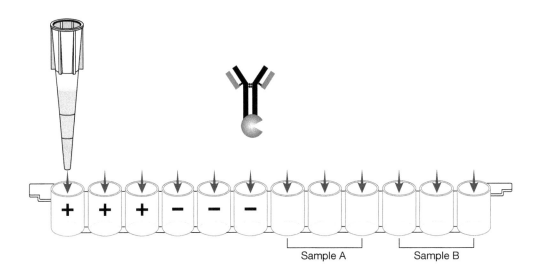

14. **Incubate 5 minutes wait 5 minutes for the antibodies to bind to their targets.**

15. **Repeat wash steps 6, 7 and 8 –THREE times**

16. **Use a fresh pipet tip to Add 50 μL "Sub" (HRP enzyme substrate) into all 12 wells.**

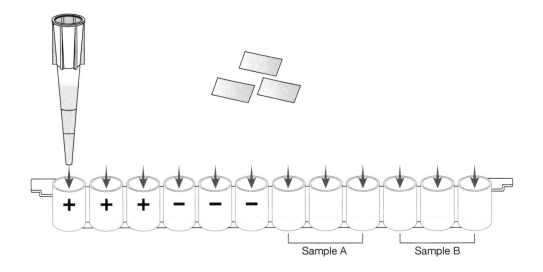

17. **Incubate for 5 minutes and watch for color change! Observe and record results on question #1 of lab answer sheet.**

18. **Deducing who was originally infected:**

    a. **Everyone Stand UP!**

    b. **If your ELISA was negative sit down**

        i. **Round 1**

            1. **If someone you shared with just sat down, also sit down**

        ii. **Round 2**

            1. **If someone you shared with just sat down, also sit down**

        iii. **Round 3 (if needed)**

            1. **If someone you shared with just sat down, also sit down**

    c. **Who is left?**

Concepts:

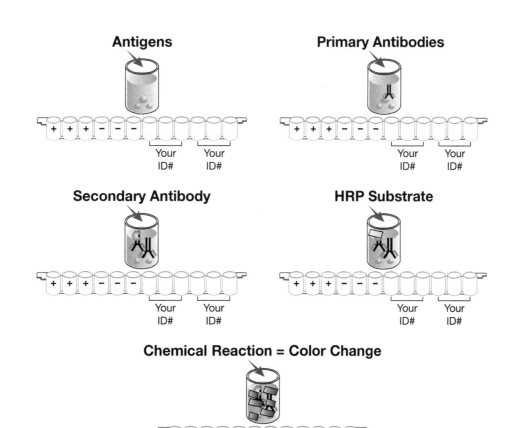

Lab 11: Pre-Lab Questions

Name: _____

Read through the entire lab and answer these questions before coming to class.

1. In this lab the antigen is describing the:
   a. Virus
   b. Antibody
   c. Enzyme
   d. Buffer

2. The first step of this lab is to
   a. Add antigen to the + wells
   b. Tap the well strip on paper towels
   c. Swap fluids with another student
   d. Look for a color change

3. It is important to use:
   a. The same pipet tip every time
   b. A different pipet tip every time
   c. An apron
   d. Sterile wells

4. A color change will only occur if:
   a. Your sample has the antibody
   b. Your sample has the antigen
   c. Your sample has the anti-serum
   d. Your sample is a negative control

5. The last steps of this lab includes
   a. Adding antigen to the + wells
   b. Tapping the well strip on paper towels
   c. Swapping fluids with another student
   d. Looking for a color change

Answer Sheet

Lab 11: ELISA

Name: _____ **Section:** _____

1. Lab Partner A Name and ID # _____

   Were you infected?   Yes   No

   Lab Partner B Name and ID # _____

   Were you infected?   Yes   No

2. Who was patient zero_____

   How do you know?_____

   _____

   _____

3. Place the following reactants in their proper order for the ELISA test:

   1 = primary antibody
   2 = secondary antibody
   3 = patient serum
   4 = substrate

   a. 2 4 1 3
   b. 3 1 2 4
   c. 1 4 3 2
   d. 4 1 3 2

4. In the ELISA test the development of color means the patient has the antibody being tested for.

   a. True
   b. False

Laboratory 12

Organs, Cavities, and Digestion

Objectives

- Locate the primary cavities of the body
- Identify and locate the organs in situ on the pig/mink and human torso model
  Quiz at end of class
- Identify and locate the components of the Digestive system from oral cavity to anus

Safety precautions: Sharp objects and toxic fluids.

Introduction

Body Regions:

- Axial region = head, neck and trunk
  - Cervical (neck)
  - thoracic (chest)
  - Abdomen
  - Inguinal
- Appendicular region = upper and lower limbs
  - upper limb (extremity) = brachial (arm), antebrachial (forearm), carpal (wrist), and digits (fingers)
  - lower limb (extremity) = femoral (thigh), crural (leg), tarsal (ankle), pedal (foot) and digits (toes)

Body Cavities and Membranes:

Spaces within the body that separates, protects, and supports internal organs.

- Major body cavities:
  - **Dorsal body cavity** (lies posteriorly)
    - Lined by **Meninges** (protective tissue membrane)
    - *cranial cavity*
    - *vertebral (spinal) canal*

- **Ventral body cavity**
  - *thoracic cavity* (encased by ribs)
    - **Pleural cavities** fluid filled area that holds the lungs
    - **Pericardial cavity** fluid filled area that surrounds the heart
  - **diaphragm muscle** separates them
  - *abdominopelvic cavity*
    - abdominal cavity
    - pelvic cavity
  - Lined by **(peritoneal) membranes**
  - Filled with viscera (organs)
    - Abdominal cavity contains: Stomach, Liver, Gallbladder, Pancreas, Spleen, Kidneys, Small Intestines, Part of Large Intestines
    - Pelvic cavity contains: Part of Large Intestines, Rectum, Urinary bladder, Reproductive organs

Digestive System (Gastrointestinal Tract {GI Tract}):

- **Primary Organs:**
  - Mouth, Pharynx, Esophagus, Stomach, Small Intestine, Large Intestine, Anus
- **Accessory organs:**
  - Teeth, Salivary glands, Tongue, Liver, Gallbladder, Pancreas
- **Digestive Functions and Processes:**
  - Ingestion: Taking food and liquid in (eating)
  - Secretion: Water, acid, buffers, hormones, and enzymes.
  - Motility (Mixing and Propulsion): Moving contents along the tube
  - Digestion: Breaking down molecules into a size that is usable by the cells.
  - Mechanical Digestion: Teeth cut food, Smooth muscles further mix and soften the food.
  - Chemical Digestion: Break down of food by enzymes. Series of hydrolysis reactions: Polysaccharides into Monosaccharides, Proteins into Amino Acids, Fats into Monoglycerides and Fatty Acids, nucleic acids into nucleotides.
  - Absorption: The passage of molecules into the cells via membrane transport (active transport, facilitated diffusion, etc); then the transfer of these molecules into the blood or lymph to be circulated throughout the body.
  - Compaction: Absorption of water and creation of feces.
  - Defecation: The excretion of material and wastes not used by the body.
  - Some nutrients are present in usable form and absorbed without being digested (vitamins, free amino acids, minerals, cholesterol and water)

Laboratory 12: Organs, Cavities, and Digestion 123

The Digestive System

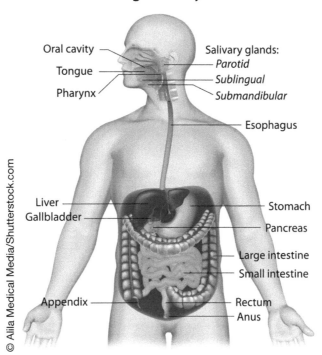

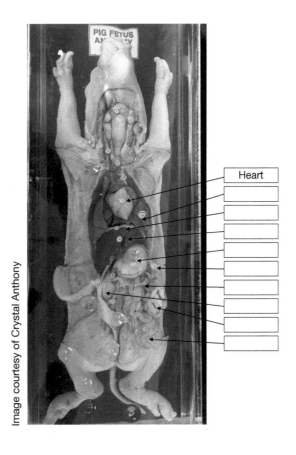

Lab 12: Pre-Lab Questions

Name: _____

Read through the entire lab and answer these questions before coming to class.

1. The body region that includes the cervical and inguinal regions is referred to as the _____ region.
   a. Axial
   b. Appendicular
   c. Cranial
   d. Spinal

2. The meninges would be found in the _____ cavity.
   a. Abdominopelvic
   b. Pleural
   c. Ventral
   d. Dorsal

3. The thoracic cavity is within the larger _____ cavity.
   a. Abdominopelvic
   b. Pleural
   c. Ventral
   d. Dorsal

4. The stomach and spleen are found in the _____ cavity.
   a. Abdominopelvic
   b. Pleural
   c. Ventral
   d. Dorsal

5. The process of taking food and liquid in is called:
   a. Ingestion
   b. Digestion
   c. Absorption
   d. Secretion

6. The process of molecules being taken up into the cells and transferred to the blood or lymph is:
   a. Ingestion
   b. Digestion
   c. Absorption
   d. Secretion

7. _____ is an example of a primary organ of the digestive system.
   a. Diaphragm
   b. Liver
   c. Esophagus
   d. Bladder

8. _____ is an example of an accessory organ of the digestive system.
   a. Diaphragm
   b. Liver
   c. Esophagus
   d. Bladder

Answer Sheet

Lab 12: Organs, Cavities, and Digestion

Name: _____   **Section:** _____

For this lab you will be using the torso model and computer resources.

The pig/mink is used to study human anatomy because, with few exceptions, the body plan of a pig/mink closely compares to humans.

1. Define the following terms

    Ventral/Anterior _____

    Dorsal/Posterior _____

    Midsagittal _____

    Parasagittal _____

    Superior _____

    Inferior _____

    Medial _____

    Lateral _____

2. There are a limited number of pigs or minks available for all of the sections. Some of them have already been dissected and some have not. If dissecting, you will need the following: Human Torso Model, Fetal Pig or Mink, Dissection Tray, Dissection kit, (2) One foot lengths of string. If it has been previously dissected, then you only need to tie back the forelimbs for observation. If needed, you may further dissect it for a more detailed view of certain anatomic features.

    Note: Only follow these instructions if dissecting a "new" pig or mink.

    1. Use the string to tie back the forelimbs and hind limbs of the pig, ventral side up, by running the string under the tray to maintain tension. Ask your instructor if you have difficulty.
    2. Using scissors, take a small snip of skin from the inferior belly region.
    3. Using the scissors, cut a midsagittal line from the inferior portion of the belly to the base of the sternum. Keep the scissors as shallow as possible to avoid damaging the internal organs.
    4. At the base of the sternum move the scissors slightly lateral and continue to make a parasagittal cut; using more force to cut through the rib cage to expose the heart and lungs.
    5. At the point of the shoulder, cut laterally to expose and open the chest cavity. Repeat this just below the rib cage. Be especially careful not to damage the organs underneath because the skin in this area is thin.

6. You can then cut diagonally along the legs at the pelvic/groin region to fully expose and create a flap at the abdominopelvic region.
7. Rinse any blood found in the body cavities at the sink. If there is any fluid in the cavity use copious amounts of water to rinse the pig. Any volume of preservatives should be "chased down" with ten times the volume of tap water.

3. Below is a list of different organs and structures you are expected to locate and learn. Write one function for each of the specified organs. Please remember to keep the functions short and simple. Your quiz at the end of class will be on this and the digestive system.

Function

Stomach _____

Liver _____

Gall Bladder _____

Pancreas _____

Spleen _____

Lungs _____

Kidneys _____

Small Intestines _____

Large Intestines _____

Heart _____

Pericardial Sac _____

Esophagus _____

Diaphragm _____

4. Describe the pathway of food from the oral cavity to the anus, using appropriate terms and anatomy:

_____

_____

_____

_____

Laboratory 13

Respiratory System

Name: _____   Section: _____

Use the words in the box to fill in the blanks. *Adapted from Lanternfish ESL worksheet.

air	oxygen	inhale	exhale
carbon dioxide	trachea	respiratory	cough
oral cavity	diaphragm	bronchi	pharynx
epiglottis	sneeze	water vapor	nasal cavity
	alveoli	blood	bronchioles

All animals need _____ to make energy from food. We get this oxygen from the _____ that we breathe. In order to get the oxygen into the blood where it can be transported to the rest of the body, the air travels through a system of organs called the _____ _____ system.

When you _____, air enters the body through the _____ or the _____. From there it passes through the _____, and then into the _____, where the _____ is located (which prevents food from entering this tube). The air travels down the trachea into two branching tubes called _____ and then on into the _____.

At the _____, oxygen from the air enters the _____. At the same time, the waste gas, _____, leaves the blood and then leaves the body when you _____. Some _____ also leaves the body when you exhale, which is why mirrors get foggy when you breathe on them. The _____ is the primary muscle that controls the lungs.

It is important to keep the respiratory system clear so oxygen can keep flowing into your body. If something gets in your nose and irritates it, you _____. If something gets in your trachea or bronchi and irritates it, you _____.

Respiratory System Worksheet

1. Label the diagram of the respiratory system below:

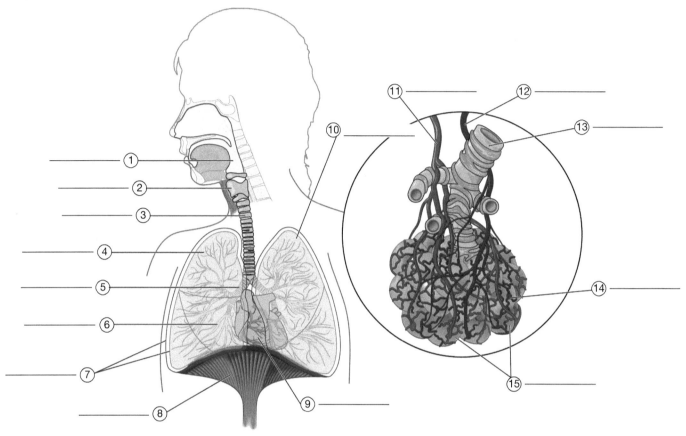

© Blamb/Shutterstock.com

2. List these structures <u>in the order in which air passes them</u> as it travels <u>from the nose to the lungs</u>. Alveoli, bronchi, bronchioles, larynx, pharynx, trachea

3. How are dust particles that enter the respiratory system in the air expelled?

4. Arrange these statements in the right order to describe inspiration (inhalation).

   ___ Air is drawn into the lungs
   ___ Air pressure in the chest cavity decreases
   ___ Lungs expand
   ___ The diaphragm contracts and flattens
   ___ The muscles between the ribs contract to move the ribs up and out

5. Add the correct term from the list below to the following descriptions: Diaphragm, Pharynx, Expiration, Alveoli, Trachea, Epiglottis, Palate, Bronchioles

   a. Smallest respiratory passageways _____

   b. Separates mouth from nose _____

   c. Windpipe _____

   d. Where gas exchange takes place _____

   e. Stops food from going the wrong way _____

   f. Both air and food pass through this _____

   g. The movement of air out of the lungs _____

   h. The main muscle involved in inspiration _____

6. Add the following labels to the diagram of the lung alveolus, below:

   <u>Alveolus</u>, Area of high <u>oxygen</u> concentration, Area of high <u>carbon dioxide</u> concentration, Movement of <u>oxygen</u>, Movement of <u>carbon dioxide</u>

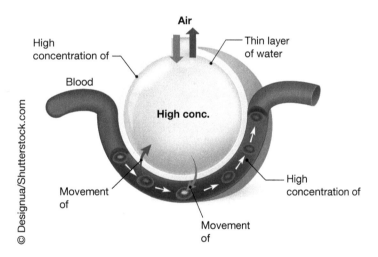

Laboratory 14

Respiratory Response

Objectives

- Use the scientific method to determine the effects of exercise on breathing
- Understand how the CNS controls breathing
- Understand how the respiratory system responds to changes in blood gas levels

Introduction

Respiration involves the inhalation and exhalation of air. The human body responds to changes in pressure and chemicals present in the blood. Stretch receptors in the carotid body and aortic body of the vessels near the heart, sense these pressure changes. This information then gets sent to the brain. The Central Nervous System (CNS) {the location of the breathing center in areas of the brain called the pons and medulla} respond accordingly. Through feedback systems, the body will maintain its baseline values by initiating responses that will bring things back to normal. For example, if the pressure in the blood vessels is too high, the CNS, will stimulate the blood vessels to relax, thereby increasing the diameter of the vessels and lowering the pressure on those vessels. Other changes such as adjustment of the heart rate can also occur. The brain will also respond to changes in chemicals present in the blood. One of these is pH, which is directly related to the amount of CO_2 in the blood. An increase in CO_2 stimulates breathing and a decrease in CO_2 inhibits it. This is why, when someone is hyperventilating, breathing in a paper bag can help them reset their breathing pattern.

Tidal Volume is the total amount of air, in Liters, inhaled and exhaled, in one breath. In the lab, you will be measuring the tidal volume as your dependent variable. You will also be measuring time, to determine the breaths per minute.

One student will be the subject of the experiment. The independent variable will be based on three conditions: hyperventilation, hypoventilation, and exercise. Each of these conditions will cause a change in CO_2 levels. You will be using a Vernier System to collect your data. Follow the procedure step by step in order to ensure your data is collected properly.

Choose which member of your group will be the subject. The subject must be a nonsmoker, with no respiratory disorder or illness. Recall that by selecting for the exclusion of these variables, we are controlling these variables. This is important because we want to exclude any variables that may impact the results of our study.

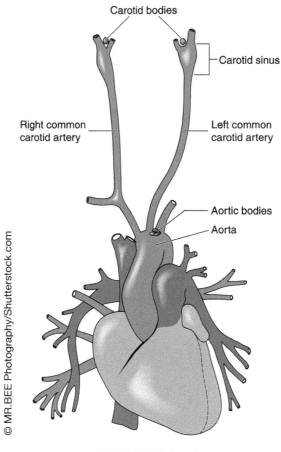

FIGURE 14.1

Materials

Vernier handheld system

Vernier Spirometer

Nose Clip

Disposable mouthpiece

Disposable bacterial filter

Procedure

1. Turn on the LabQest machine. Plug it in if needed. Connect the Spirometer, then choose New from the File menu.
2. On the Meter screen, tap Rate. Change the data-collection rate to 100 samples per second and the duration to 120 seconds. Select OK. You must do this again after every run. At this setting, the graph will collect data for 120 seconds. You do not have to go for the entire 120 seconds. Once you complete your postchallenge breaths you may click the stop button ■.

3. First, attach the disposable bacterial filter to the side of the spirometer with "Inlet" written on it. Then, attach a disposable spirometer mouthpiece to the other end of the bacterial filter (Figure 14.2).

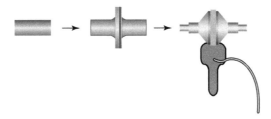

FIGURE 14.2

4. Have the student who is the subject hold the wrapping their lips around the cardboard mouthpiece. Choose Zero from the Sensors menu.

5. Process for collecting the data:
   a. Put on the nose clip.
   b. Press the Play button ▶. This starts the recording of breath data on the graph.

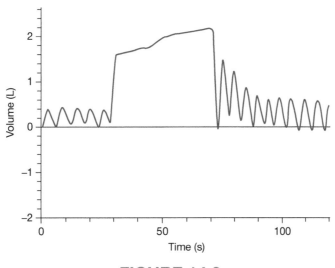

FIGURE 14.3

   c. Take 4 normal breaths for the prechallenge data. Continue recording as you do the challenge (hypoventilation/breath holding). You will fill our lungs as deeply as possible (maximum inspiration) and try to hold your breath for 20 to 40 seconds. Then breath normally again, for at least 4 breaths. This is your postchallenge data.
   d. This one run will give you all the data you need for your challenge.

6. Tap the left side of the graph (the y-axis) to show volume in liters. Time is on the x-axis in seconds. View a graph of volume versus time, tap the y-axis label and select Volume. Zero is the baseline, anything above that line is a positive number and anything below that line is a negative number. If needed, you can adjust the baseline. To do this, select Analyze Menu, then Advanced, then Baseline Adjustment and then ok.

7. Record the data in table 1. From one valley to the next valley is one full breath cycle. Choose one breath cycle to measure. Use the stylus to click at the top of one peak in the Pre-Challenge portion of the data. Note the volume and write it in the table under Peak Volume. Then, click on the corresponding valley, just to the right of the peak you just measured. Note the volume and write it in the table under Valley Volume. If this number is negative, write it as a positive number. Add these numbers together to get Tidal Volume (Total Volume).

Peak Volume + Valley Volume = Total Volume (Tidal Volume).

At this time you can also note the time in seconds (of the valley). Record this time in seconds, on the data table, in the 2nd Valley column. Finally, put the stylus at the valley just before the peak and record this time in the 1st Valley column of the data table. Subtract the 1st Valley time from 2nd Valley time to get the time of one breath cycle.

2nd Valley − 1st Valley = Total Difference (Time of one breath cycle)

Now calculate the breaths per minute by dividing 60 seconds by the total difference.

$$\frac{60 \text{ seconds}}{\text{Total difference}} = \text{breaths per minute}$$

8. Repeat step 7 for the Postchallenge portion of the data.
9. Repeat the procedure for Hyperventilation/fast breathing. Instead of holding your breath, you will breathe deeply and rapidly. If you get lightheaded slow down your breathing.
10. Repeat the procedure for the exercise condition. Instead of holding your breath or fast breathing, you will perform exercise. Jogging in place or jumping jacks will work.

Answer Sheet

Lab 14: Respiratory Response

Name: _____ **Section:** _____

TABLE 1 Data

	Peak Volume (L)	Valley Volume (L)	Tidal Volume Total (L)	2nd valley (seconds)	1st valley (seconds)	Total Difference	Respiratory Rate Breaths/min
Hypoventilation							
Pre-Challenge							
Post-Challenge							
Hyperventilation							
Pre-Challenge							
Post-Challenge							

TABLE 2 Exercise

	Peak Volume (L)	Valley Volume (L)	Tidal Volume Total (L)	2nd Valley (seconds)	1st Valley (seconds)	Total Difference (seconds)	Respiratory Rate (breaths/min)
Prechallenge							
During Challenge							
Postchallenge							

End of Lab Questions

1. Explain what happened to the Tidal Volume after each condition (hyperventilation, hypoventilation, and exercise).

2. Explain what happened to the Respiratory Rate after each condition (hyperventilation, hypoventilation, and exercise).

3. Which condition resulted in the biggest pre to post challenge changes when comparing all conditions?

4. Compare the pre and post data for each condition and then compare them to the other groups/conditions. What conclusions can you make based on the data you collected?

5. Based on what you learned in this lab, what advice would you give to someone who wanted to improve their fitness?

Laboratory 15

Urinalysis

Objectives

- Define and perform the steps involved in formulating a medical diagnosis
- Explain the difference between signs and symptoms
- Make observations and perform tests to determine the presence of substances in a urine sample

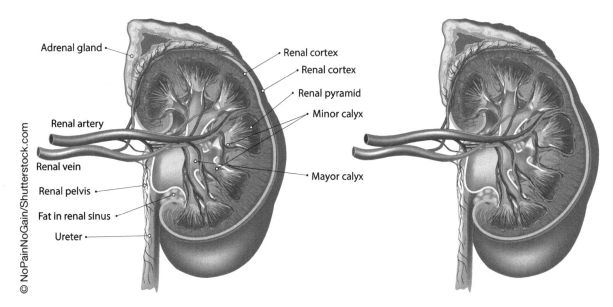

FIGURE 15.1

Introduction

The renal system includes the kidneys and the ureters which lead to the urinary bladder. The kidneys produce urine which is then sent to the urinary bladder for storage. Kidneys act as a filtration system. Toxins and wastes are processed and sent into the urine for excretion and elimination from the body. Analyzing the urine is a good way to diagnosis a patient. The medical community has been able to determine signs that indicate certain conditions. Sometimes there are substances present in urine, such as sugars, protein and crystals, that should not be there at all. With certain conditions, such a Renal Failure, protein may appear in the urine as clumps, called protein casts. Color and pH can also give indications of certain disorders. Some cells are

137

found in normal urine. These cells are surface layer epithelial cells that have shed. Other cells, such as white blood cells and red blood cells may be found and can indicate an infection or kidney tissue damage, respectively.

Signs are what the health professional observes or measures. Symptoms are what the patient reports. It is a combination of all of the information the health professional gathers that helps them to make a diagnosis. You will be following the procedures to analyze four urine samples. Table 15.1 provides information regarding the patient in which the sample can from. You will compare the data you collect with the information in the tables.

TABLE 15.1 Patient Background Information

PATIENT	AGE	
#1	22	Polyuria (excess urination), polyphagia (excessive appetite), polydipsia (excessive thirst) and unexplained weight loss
#2	65	Dizzy, trouble sleeping, swollen ankles, pain in mid back
#3	28	Dysuria (painful urination), Polyuria (excess urination), fever which resolves with antibiotics
#4	NA	Normal – control sample

Lab 15: Pre-Lab Questions

Name: _____

1. Protein casts are:
   a. Always present in urine
   b. Sometimes present in urine, but it is normal
   c. Sometimes present in urine, but indicates a disorder
   d. Never present in urine

2. A red color is:
   a. Always present in urine
   b. Sometimes present in urine, but it is normal
   c. Sometimes present in urine, but indicates a disorder
   d. Never present in urine

3. An urine sample has a pH of 9, this is:
   a. Always the pH of urine
   b. Sometimes the pH of urine, but it is normal
   c. Sometimes the pH of urine, but indicates a disorder
   d. Never the pH of urine

4. Cells are:
   a. Sometimes present in urine, but may indicate a disorder
   b. Sometimes present in urine, but it is normal
   c. Never present in urine
   d. Is abnormal and indicates a disorder

5. In medicine, signs are:
   a. What is measured by the medical professional
   b. What is reported by the patient
   c. What tells you which way to turn
   d. Is not relevant to the diagnosis

Procedures

1. To test for color:
   a. Label each plastic cup with the patient number.
   b. Shake the urine sample and use the pipet to transfer 10 mL of the sample into the correct cup.
   c. Look at color and record it in the data table.

TABLE 15.2 Color

Sign	Indication
YELLOW	Normal
PALE YELLOW	Diabetes insipidus, Granular kidney, Normal for lots of water
MILKY/CLOUDY	Urogenital Tract Infection (UTI), Normal for fat globules
AMBER/RED	Renal Tissue Damage, UTI, Cancer (Blood), Normal for certain medications or eating beets
GREEN	Jaundice (Bile), Some Bacterial Infections
BROWN /BLACK	Phenol or Metal Poisoning, Hemorrhage (renal injury, malaria)

2. To test for pH:
   a. Use a pH strip to test the sample.
   b. Use a different pH strip for each sample.
   c. Compare the pH strip to the chart to determine the pH
   d. Record the pH in the data table.

TABLE 15.3 pH

Sign	Indication
6 to 8	Normal
ACIDIC	Fever, Acidosis
ALKALINE	Anemia, Vomiting, Ischuria (Urine Retention), UTI

3. To test for Sugar:
   a. Use pipets to accurately measure 3 mL of urine sample and put it into a test tube. Use a new pipet for each sample.
   b. Add 3 mL of Benedict's solution to the test tube with the urine in it. Use a new pipet for each sample.
   c. Heat for 10 minutes.
   d. Record the color in the data table.

e. Negative = blue/green. Positive = Copper Precipitate; see examples on display if available.
f. Repeat for all of the samples. Remember to use new pipets.
g. Poor contents into disposal bin and wash test tubes with soap and water, using test tube brush.

4. To test for Protein:
   a. Use pipets to accurately measure 3 mL of urine sample and put it into a test tube. Use a new pipet for each sample.
   b. Add 1 ml of Biuret solution to the test tube with the urine. Use a new pipet for each sample.
   c. Swirl the tube.
   d. Record the color in the data table.
   e. Negative = blue. Positive = Red/Purple.
   f. Repeat for all of the samples.
   g. Poor contents into disposal bin and wash test tubes with soap and water, using test tube brush.

TABLE 15.4 Molecules

Sign	Indication
PROTEIN	Kidney disorder, renal tissue damage, chronic kidney failure
GLUCOSE	Diabetes mellitus
KETONES	Low carb diet/excess fat metabolism, diabetes mellitus

5. To find cells, crystals, and protein casts
   a. View the microscope slide for each patient.
   b. Look throughout the entire sample for Erythrocytes (red blood cells), Leukocytes (white blood cells), crystals and/or protein casts.
   c. Record your findings in the data table.
   d. Repeat for each patient.

TABLE 15.5 Cells, Crystals and Protein Casts

Sign	Indication
RED BLOOD CELLS	Renal tissue damage/chronic kidney failure, UTI, cancer
WHITE BLOOD CELLS	Infection
CRYSTALS	Kidney stones, UTI, diabetes mellitus, gout, chronic nephritis
PROTEIN CASTS	Chronic kidney failure

Answer Sheet

Lab 15: Urinalysis

Name:_____   Section:_____

DATA TABLE 15.1 Color and pH

Patient	Color	pH
#1		
#2		
#3		
#4		

DATA TABLE 15.2 Sugar

Patient	Color	Positive or Negative
#1		
#2		
#3		
#4		

DATA TABLE 15.3 Protein

Patient	Color	Positive or Negative
#1		
#2		
#3		
#4		

DATA TABLE 15.4 Cells, Crystals, and Protein Casts

Patient	Red Cells	White Cells	Crystals	Protein Casts
#1				
#2				
#3				
#4				

For each of the patients tell what disorder or disease you think the patient has. Tell all the signs and symptoms each patient has. Be specific and describe what factors led you to this decision. For patient number 4, describe the normal features of urine and why it is important to have a control sample.

Patient #1

Patient #2

Patient #3

Patient #4

Laboratory 16

Skeletal System – Bone Lab

Objectives

- Identify the major bones of the human body **Quiz at end of class**
- Describe what bones and regions make up the Axial Skeleton versus the Appendicular Skeleton

Introduction

You will be using the bone box and computer resources to learn the following list of bones. You are required to be able to identify all of the names of the bones and the parts of the bones listed in this lab.

Axial Skeleton

Skull

Frontal Bone

Parietal Bone (2)

Temporal Bone (2)

Occipital Bone

Sphenoid Bone

Nasal Bone

Zygomatic Bone (2)

Maxilla

Mandible

Hyoid Bone

Spinal Column

Parts of typical vertebrae: Body, Spinous Process, Transverse Process, Vertebral Canal

Cervical Vertebrae (7)

Thoracic Vertebrae (12)

Lumbar Vertebrae (5)

Sacrum (5 fused vertebrae)

Coccyx (4 fused vertebrae)

Intervertebral Disc

Rib Cage

Ribs (12)

True Ribs (7)

False Ribs (8 to 12) (Floating ribs 11 and 12)

Costal Cartilage

Sternum: Manubrium, Body, Xiphoid Process

Appendicular Skeleton

Upper Limb (Extremity)

Clavicle (2)

Scapula (2): Glenoid Cavity, Coracoid Process, Acromion Process

Humerus (2): Greater Tubercle, Bicipital Groove, Epicondyles (Lateral/Medial)

Ulna (2): Olecranon, Head, Trochlear Notch

Radius (2): Head, Styloid Process

Carpals (8): Scaphoid, Lunate, Triquetrum, Pisiform, Trapezium, Trapezoid, Capitate, Hamate

Metacarpals (5): I, II, III, IV, V

Phalanges (14): Proximal, Middle, Distal / I, II, III, IV, V

Lower Limb (Extremity)

Pubic (Coxal) Bone (2): Ilium, Ischium, Pubis; Acetabulum

Femur (2): Head, Greater Trochanter, Condyles (Lateral/Medial)

Patella

Fibula: Head, Lateral Malleolus

Tibia: Tibial Plateau, Tibial Tuberosity, Medial Malleolus

Tarsal Bones (6): Calcaneus, Talus, Navicular, Cuboid, 3 Cuneiforms

Metatarsals (5): I, II, III, IV, V

Phalanges (14): Proximal, Middle, Distal / I, II, III, IV, V

Lab 16: Pre-Lab Questions

Name: _____

Read through the entire lab and reference your book to answer these questions before coming to class.

1. The skeletal region that includes the cervical and thoracic regions is referred to as the _____ skeleton.
   a. Axial
   b. Appendicular
   c. Coxal
   d. Extremity

2. The occipital bone is part of the _____.
   a. Upper Limb
   b. Lower Limb
   c. Skull
   d. Spinal Column

3. The vertebral bone is part of the _____.
   a. Upper Limb
   b. Lower Limb
   c. Skull
   d. Spinal Column

4. The spinous process is part of the _____.
   a. Vertebrae
   b. Sacrum
   c. Coccyx
   d. Sternum

5. The greater trochanter is part of the _____.
   a. Humerus
   b. Radius
   c. Femur
   d. Tibia

6. The styloid process is part of the _____.
   a. Humerus
   b. Radius
   c. Femur
   d. Tibia

7. The tarsal bones include all of the following, EXCEPT.
   a. Calcaneus
   b. Tarsal
   c. Cuboid
   d. Hamate

8. _____ is an example of a carpal bone.
   a. Calcaneus
   b. Tarsal
   c. Cuboid
   d. Hamate

Laboratory 17

Muscular System: Skeletal Muscles Lab

Objective

- Identify the Major skeletal muscles of the human body **Quiz at end of Class**

Introduction

You will be using the muscle man models and computer resources to learn the following list of muscles. You may also be viewing and/or dissecting a pig or mink. You are required to be able to identify all of the names of the muscles listed in this lab.

Anterior	Posterior
Face	
Orbicularis Oculi Masseter Orbicularis Oris	
Trunk	
Sternocleidomastoid (neck) Pectoralis Major Serratus Anterior External Oblique Rectus Abdominis	Trapezius Latissimus Dorsi (thoracolumbar fascia)
Upper Extremity	
*Deltoid *Biceps Brachii *Brachioradialis *Flexor Carpi Group	*Triceps Brachii *Extensor Carpi Group

(continue)

Lower Extremity	
*Adductors (Longus/Magnus) *Sartorius *Gracilis *Quadriceps Group (Rectus Femoris/ Vastus Intermedius/Vastus Lateralis/ Quadriceps Femoris) *Tibialis Anterior *Extensor Digitorum Longus	*Tensor Fascia Latae Gluteus (Maximus/Medius) *Hamstrings Group (Biceps Femoris/ Semitendinosus/ Semimembranosus) *Gastrocnemius

*=identify on pig/mink

If dissecting you will need the following:

Muscle Man Model, Fetal pig or Mink, Gloves, Dissection tray, Dissection kit, (2) One foot lengths of string

Safety precautions: Sharp objects and toxic fluids.

Procedure

Most classes will be instructed to dissect only one extremity. There are a limited number of pigs/minks available for all of the sections. Some of the pigs/minks have already been dissected and some have not. If the pig has been previously dissected, then you will do a limb that has not yet been dissected. If it is completely dissected, then you can clean up and tidy up muscles or just go straight to learning to identify them.

Ask your instructor if you have difficulty at any point.

1. Use the string to tie back the forelimbs and hind limbs of the pig/mink, by running the string under the tray to maintain tension.
2. If there is skin, take a small snip of skin from the extremity you have been instructed to dissect.
3. Using the scalpel, cut a line along the lateral and posterior extremity. Keep as shallow as possible to avoid damaging the muscles. Fetal pig muscles and Mink muscles are very small. The skin is thin and there is not much fat.
4. Use fingers or forceps to pull back and separate skin. Use scalpel to cut away fascia and fat.
5. Clean up muscles by removing fat and fascia that surrounds to muscles. You will not likely need a scalpel. You can primarily use forceps, probes and scissors. This will minimize damage to the specimen and decrease chances of injury to students.
6. Locate and identify the muscles of your extremity.
7. Locate and identify all the muscles on the muscle list. You are responsible for identifying all the muscles on diagrams, models and specimens for the lab practical.

Note: If the entire pig/mink has been dissected, then you may only need to tie back the forelimbs for observation. If needed, you may further dissected the pig/mink for a more detailed view of certain anatomic features.

Lab 17: Pre-Lab Questions

Name: _____

Read through the entire lab and reference your book to answer these questions before coming to class.

1. The region that includes the Orbicularis Oculi muscle is the _____.
   a. Face
   b. Trunk
   c. Upper Extremity
   d. Lower Extremity

2. The region that includes the Gracilis muscle is the _____.
   a. Face
   b. Trunk
   c. Upper Extremity
   d. Lower Extremity

3. The region that includes the Deltoid muscle is the _____.
   a. Face
   b. Trunk
   c. Upper Extremity
   d. Lower Extremity

4. The region that includes the Serratus Anterior muscle is the _____.
   a. Face
   b. Trunk
   c. Upper Extremity
   d. Lower Extremity

5. The Quadriceps muscle group is located _____.
   a. Anterior to the humerus
   b. Posterior to the humerus
   c. Anterior to the Femur
   d. Posterior to the Femur

6. The Pectoralis muscle is located _____.
   a. On the back
   b. On the chest
   c. On the face
   d. On the leg

7. The Tibialis Anterior muscle is located _____.
   a. Anterior to the tibia
   b. Posterior to the tibia
   c. Anterior to the ulna
   d. Posterior to the ulna

8. The Biceps muscle group is located _____.
   a. Anterior to the humerus
   b. Posterior to the humerus
   c. Anterior to the Femur
   d. Posterior to the Femur

Laboratory 18

Brain Models and Dissection

Objectives

- Identify and locate the basic anatomy of the brain and spinal cord ***Quiz at end of class***
- Understand the basic functions of the brain regions
- Understand the relationship of nervous system regions
- Safely and correctly use dissection techniques

Safety precautions: Sharp objects and toxic fluids.

Introduction

Dissection helps students to learn, identify and visualize anatomical structures. Comparative anatomy is useful because animals have structures similar to humans. We will be using sheep or cow brain and models to learn the anatomy of the heart. You will follow the instructor's instructions for cutting open the brain.

You will also be using computer resources and picture mats to help you learn the parts of the brain. You are required to be able to identify all of the parts of the brain and spinal cord listed in this lab.

Locate and identify the following structures on the internal and external surfaces of the brain:

Cerebrum
- Lobes: Frontal, Parietal, Occipital, Temporal
- Left and Right Hemispheres
- Sulci and Gyri
- Medial Longitudinal Fissure

Corpus Callosum

Mid Brain
- Thalamus
- Hypothalamus

Optic Tract/Optic Nerve/Optic Chiasm

Brain Stem

- Midbrain
- Pons
- Medulla

Cerebellum

Spinal Cord

White and Grey Matter

Identify the following structures on the diagram/model of the spinal cord:

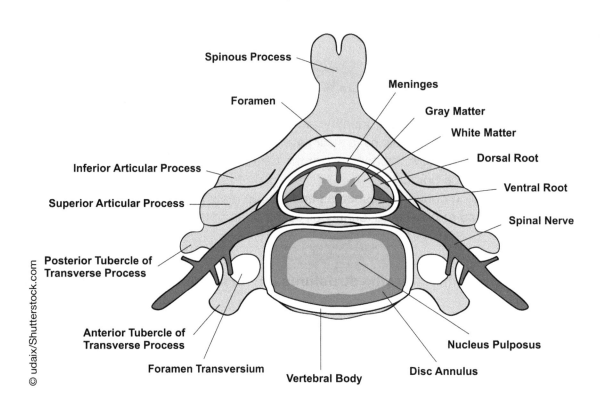

Lab 18: Pre-Lab Questions

Name: _____

Read through the entire lab and reference your book to answer these questions before coming to class.

1. The region that includes the Hypothalamus is the _____.
   a. Cerebrum
   b. Mid Brain
   c. Brain Stem
   d. Spinal Cord

2. The region that includes the Parietal Lobe is the _____.
   a. Cerebrum
   b. Mid Brain
   c. Brain Stem
   d. Spinal Cord

3. The region that includes the Pons is the _____.
   a. Cerebrum
   b. Mid Brain
   c. Brain Stem
   d. Spinal Cord

4. The region that includes the Dorsal Root Ganglion is the _____.
   a. Cerebrum
   b. Mid Brain
   c. Brain Stem
   d. Spinal Cord

5. The region that includes the Medial Longitudinal Fissure is the _____.
   a. Cerebrum
   b. Mid Brain
   c. Brain Stem
   d. Spinal Cord

6. The optic tract is in the _____ region of the brain.
   a. Anterior
   b. Lateral
   c. Posterior
   d. Proximal

7. The cerebellum is in the _____ region of the brain.
   a. Anterior
   b. Lateral
   c. Posterior
   d. Proximal

8. The white and grey matter is _____.
   a. Not part of the brain
   b. Only in the cerebrum
   c. Only in the spinal cord
   d. None of the above

Laboratory 19

Introduction to EMG

Objectives

In this experiment, you will

- Obtain graphical representation of the electrical activity of a muscle.
- Associate amount of electrical activity with strength of muscle contraction.
- Compare wrist flexor muscle function during different types of gripping activity.

Introduction

An electromyogram, or EMG, is a graphical recording of electrical activity within muscles. Activation of muscles by nerves results in changes in ion flow across cell membranes, which generates electrical activity. This can be measured using surface electrodes placed on the skin over the muscle of interest.

Electrical activity correlates with strength of muscle contraction, and is dependent on the quantity of nerve impulses which are sent to the muscle. This is easily visible in large muscles such as the biceps muscle in the arm and the quadriceps muscle in the leg, but can also be demonstrated in smaller, less visible muscles, such as the masseter muscle in the jaw.

Tendonitis of the wrist flexor muscle group at the medial epicondyle, golfer's elbow, results from repetitive stress, poor joint alignment, and muscular imbalance at the elbow joint. Problems here primarily affect wrist and hand actions; such as gripping and wrist flexion. This affects many types of activities such as playing guitar, sewing, and cooking. It is especially debilitating and troublesome if it prevents people from doing their daily work duties (which is often the repetitive stress that led to the problem in the first place). In this experiment, you will examine the electrical activity generated by squeezing a ball of various strengths of resistance. You will see how different amount of efforts in the gripping requirements influences the strength of contraction in the forearm musculature.

Important: Do not attempt this experiment if you suffer from pain in or around the elbow or wrist. Inform your instructor of any possible health problems that might be exacerbated if you participate in this exercise.

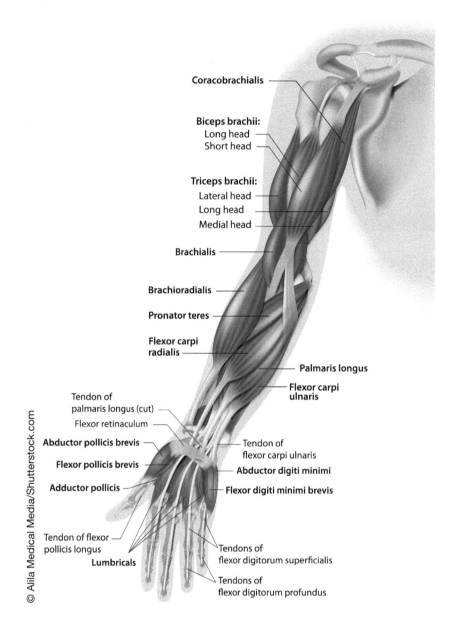

Materials

LabQuest
Electrodes
Alcohol wipes
Vernier EKG sensor
Therapy balls of varied resistance

Procedure

Part I – Conscious Clenching of the Fist

Select one person from your lab group to be the subject.

1. To minimize interference with readings, turn off cell phones and unplug LabQuest unit. Connect the EKG Sensor to LabQuest and choose New from the File menu.

2. On the Meter screen, tap Duration. Change the data-collection length to 30 seconds. Tap on Rate and put on 100 s/s. Put the interval at .01. Select OK.

3. Instruct the subject to be seated. Remove excess oil from the skin with soap and water or alcohol to improve the adhesion of the electrode tabs to the skin. Position the upper electrode tab slightly below the medial epicondyle (see picture). Position the lower tab medial and distal to the first electrode. Attach the EKG electrodes to the tabs; in this experiment, red and green leads are interchangeable. Place a third electrode tab on the mid-distal forearm and attach the black EKG electrode to this tab.

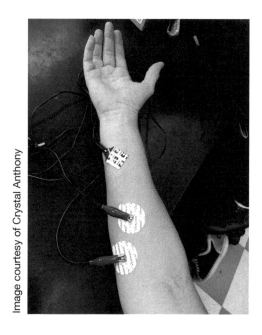

Image courtesy of Crystal Anthony

4. Have the student sit with his or her arm relaxed, fingers extended. Start data collection by pressing the play button ▶. If your graph has a stable baseline for 5 seconds (see graph), stop data collection and continue to Step 5. If your graph has an unstable baseline, stop data collection and try again until you have a stable baseline for 5 seconds. If you do not have a stable baseline adjust rate to 200 s/s and try again. Playing with the rate may help tighten the line. Makes sure electrodes are attached, not lifted or shifting.

5. Start data collection. Click on the left side (Y-axis Title) of the graph on the LabQuest device and change to Potential (as shown on graph). After recording 5 seconds of stable baseline with the arm relaxed, instruct the subject to moderately clench his/her empty fist for 5 seconds, then relax. Repeat this process of clenching for 5 seconds and relaxing for 5 seconds to obtain several events. Data collection will end after 30 seconds.

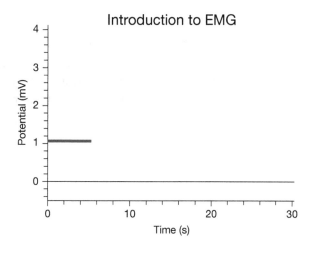

6. One the graph, find a representative peak potential during the time when the subject's arm was relaxed.
   a. Use the stylus to tap the top of the peak to find the max. value and the bottom of the peak (ie. the valley) to find the min. value. Record these values in Table 1 for Resting Interval, rounding to the nearest 0.01 mV.
   b. Another option is to click on the X/Y graph icon on the top of the screen, then tap the toggle button until you have the time period where you chose you representative peak. This will show data for that region of time and you can chose the largest and smallest values shown in that data set to record in Table 1 for Resting Interval.
   c. Record the minimum and maximum values in Table 1, rounding to the nearest 0.01 mV. You will see Min, Max and ΔY. ΔY is the change in mV, ΔmV.
   d. Choose Statistics from the Analyze menu to turn off statistics.
7. Repeat step 6, except for this step you will find a representative peak potential when the subject was clenching his/her fist.
   a. Tap and drag over a 5-second interval (a best representation of the three trials), during which the subject was clenching his/her fist.
   b. Choose Statistics from the Analyze menu.
   c. Record the minimum and maximum values in Table 1, rounding to the nearest 0.01 mV.

Part II – Comparison of Muscle Action in Gripping Different Resistance Balls

8. Repeat steps for recording releaxing and gripping intervals for each of the different colored balls.
   a. Start data collection. If your graph has a stable baseline for 5 seconds, stop data collection and continue to Step 9. If your graph has an unstable baseline, stop and try again until you have a stable baseline for 5 seconds. Use the trouble shooting ideas in Step 4.

9. Start data collection. After recording 5 seconds of stable baseline, instruct the subject to grip the ball, only using enough strength to get slight movement of the ball when squeezing. Do not use your max grip effort. Then have the subject relax his/her arm, for 5 seconds. Repeat this process of clenching for 5 seconds and relaxing for 5 seconds, to obtain several events. Data collection will end after 30 seconds. Note: If you are confident with the data collected during the gripping action, you do not have to relax and repeat the action. You are not required to record for the entire 30 seconds.

10. Refer to steps 6 and 7 to record data in Table 1.

11. Calculate ΔmV by combining the min and max data values. If the Min. Value is a negative number and the Max. Value is a negative number then subtract the two numbers to find the amplitude of contraction. If the Min. Value is a negative number and the Max. Value is a positive number, then add the two numbers to find the amplitude of contraction, (take the absolute value of the negative number; make it a positive number).

Answer Sheet

Lab 19: Introduction to EMG

Name: _____   Section: _____

Data

Table 1				
Condition	Interval	Minimum mV	Maximum mV	ΔmV
Hand gripping	Resting Interval			
	Gripping Interval			
Squeezing Ball 1 (Blue)	Resting Interval			
	Gripping interval			
Squeezing Ball 2 (Yellow)	Resting Interval			
	Gripping Interval			
Squeezing Ball 3 (Purple)	Resting Interval			
	Gripping Interval			

Data Analysis

1. Compare the values (mV) of muscle activation during gripping of the three balls tested. Is there a difference in the potential value generated during the gripping interval of data collection for each of the items tested? Explain in detail referencing your data.

2. For each colored ball, rank from greatest (1) to least (3), the EMG electrical activity for each of the resistive balls tested.

   1.

   2.

   3.

3. Compare your data with other lab groups. Are there any differences? If there are differences, why do you think you got different results?

4. Which specific muscles are activated in this experiment? What exercises and or stretches might someone do to strengthen and lengthen these muscles, so that they could successfully performed their hobby or job activity without injury?

5. What are some ways to reduce/prevent developing tendonitis?

Laboratory 20

Human Senses

Objectives

- Determine the location on the tongue where the sensations of sweet, sour, and bitter are located
- Determine how accurate a subject is at localizing the sense of touch
- Determine the difference in density of touch receptors of different parts of the skin
- Demonstrate the presence of the blind spot
- Discover the characteristics of hearing
- Investigate sensory transduction by testing reaction time.

Introduction

A property of all living systems is the ability to detect and respond to environmental factors, or **stimuli**, that are crucial to survival and/or reproduction. The purpose of this lab is to investigate the human **sensory system** and its integration with the **muscular (motor) system** to produce adaptive responses to various types of stimuli.

The nervous system works together as an integrated communication and information processing organ system. The different parts of the nervous system communicate through specialized cells called **neurons**. Neurons have four important parts: the cell body, axon, synapse, and dendrites. The **cell body** of the neuron contains the nucleus and most of the organelles. The axon is a long extension off of the cell body that transmits nerve signals to the synapse. The **synapse** connects neurons to other neurons and muscles. The **dendrites** are the receiving end of the neuron that detect signals from other neurons and transmits this information back to the cell body (Figure 20.1).

The sensory system is a part of the overall nervous system that includes the **central nervous system** (CNS) and **peripheral nervous system** (PNS). The CNS is composed of the **brain** that is the site of information processing and integration and the **spinal cord** that connects the brain to the rest of the body. The peripheral nervous system is composed of sensory neurons that bring information from the **sensory receptors** back to the brain and motor neurons that control voluntary contractions of skeletal muscles and involuntary movements of the muscles in the heart and other internal organs. Sensory information is collected by sensory receptors in sensory organs like the eye and transmitted back to the brain by sensory neurons. The brain integrates this information and commands muscle contractions in response to the sensory information.

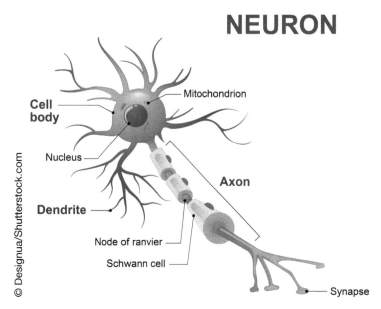

FIGURE 20.1 Neuron structure.

This process of sensory information processing and muscle responses is called sensory transduction (Figure 20.2).

Although you will be examining the various sensory receptors of the body, it should be remembered that these receptors are not where feeling is actually experienced. Instead, they are merely receptive to stimulation and transmit information to the central nervous system where the sensation of feeling is actually produced.

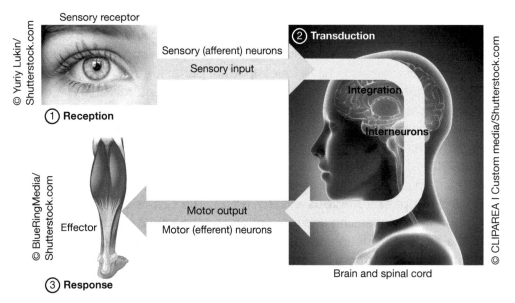

1. Sensory receptors → 2. signal transduction → 3. Physiological response

FIGURE 20.2 Sensory transduction.

Lab 20: Pre-Lab Questions

Name: _____

Read through **all** of the lab handout and answers these questions **before** coming to class.

1. The _____ is the part of the neuron that sends information to other neurons.
   a. Axon
   b. Dendrites
   c. Cell body
   d. Synapse

2. The nucleus is found in the _____ of the neuron.
   a. Axon
   b. Dendrites
   c. Cell body
   d. Synapse

3. The _____ system in the body receives external stimuli and transmits sensory information to the brain.
   a. Central nervous
   b. Spinal cord
   c. Sensory
   d. Motor output

4. Which of these is the sensory receptor for vision?
   a. Chemoreceptor
   b. Photoreceptor
   c. Mechanoreceptor
   d. Pain Receptor

5. Which of these is the sensory receptor for hearing?
   a. Chemoreceptor
   b. Photoreceptor
   c. Mechanoreceptor
   d. Pain receptor

6. Which part of the nervous system includes the brain and spinal cord?
   a. Central nervous system
   b. Peripheral nervous system
   c. Sensory system
   d. Motor output

7. Which of these is not one of the five primary human tastes?
   a. Salty
   b. Sweet
   c. Spicy
   d. Bitter

8. Which of these is the sensory receptor for touch?
   a. Chemoreceptor
   b. Photoreceptor
   c. Mechanoreceptor
   d. Pain receptor

9. What type of instrument will we be using to test hearing ability?
   a. Tuning fork
   b. Radio
   c. Whistle
   d. Horn

10. Which of these is the connection between neurons?
    a. Axon
    b. Dendrites
    c. Cell body
    d. Synapse

Procedure

Part I – Taste

Fish, primarily bottom feeders, have **chemoreceptors** scattered over the surfaces of their bodies. These cells play a major role in determining their behavior. In the terrestrial animals, taste cells are located inside the mouth where they act as sentinels, providing a final judgment on what should not be swallowed. Humans are able to distinguish five primary tastes; sweet, salty, bitter, sour, and savory (or umami). Each primary taste appears to stimulate a different type of sensory receptor called a taste bud (plus supporting cells). The five types of receptors are not evenly distributed on the tongue; some are more concentrated toward the tip, others on the sides, etc. In this exercise, you will map the sense of taste on the tongue.

1. Obtain a cotton swab and dip it into one of the taste solutions on the side of the labroom. The solutions are sweet, sour, salty and bitter (we will not be testing savory in this experiment).

2. Map the surface of the tongue by touching various regions of the tongue with the swab. Write a "+" sign on the tongue maps in the results section to show the locations where the chemical is detected. Rinse your mouth with water between each solution to cleanse your palate.

3. Dry off your tongue with a clean paper towel. Place a few grains of table salt on the tip of your tongue. Record the time interval from the time you place the salt crystals on the tip of your tongue until you first taste the salt in Table 1 in the results section.

4. Using the salt solution, place some liquid salt on the tip of your tongue. Record the time interval from the time you place the salt solution on the tip of your tongue until you first taste the salt.

Part II – Touch

The sense of touch is made up of a number of different types of **mechanoreceptor** sensory organs. These receptors are found more densely distributed in some parts of the body than in others. Pressure, pain, heat, and cold are all aspects of the sense of touch.

Localization of Touch

1. The subject should keep his or her eyes closed throughout the exercise.

2. Touch the skin on the back of the hand, the palm, or the forearm of the subject lightly with the pointed end of a washable marker. Be sure to leave a mark.

3. Touch the skin on the back of the hand, the palm, or the forearm of the subject lightly with the pointed end of a washable marker. Be sure to leave a mark.

4. Then ask the subject (with eyes still closed) to use a different color marker to locate the place on the skin where the stimulus was received.

5. Use a ruler to measure, in millimeters, as closely as possible the error in locating where the stimulus was applied. Record your measurements in Table 2 in the results section.

6. Test this location one more time (Trial #2).

7. Repeat steps 3 to 6 for the other two locations (back of the hand, palm, and forearm).

8. Change roles with your partner and repeat the experiment.

Density of Sense Organs

1. Have the subject keep his or her eyes closed.

2. Set the calipers to the first distance (20 mm) listed in the table below. Using calipers gently touch the subject's skin on the back of the hand, the palm, or the forearm so that the either a single point or both points of the instrument touch with light pressure. Your goal is to test whether your subject can discriminate one or two points so keep them guessing!

3. Ask your partner to state whether one of the two points of the instrument is felt.

4. If your subject guessed right put a "+" sign in Table 3 if she/he got it wrong put a "−" sign in Table 3.

5. Repeat this procedure 2 times for the first area chosen at the first distance of 20 mm. Record your results in Table 3.

6. Now decrease the distance between the caliper points and test the same area following steps 1 to 5 at 15, 10, and 5 mm.

7. Test the other two locations (the back of the hand, the palm, or the forearm) at all three distances

8. Change roles with your partner and repeat the experiment.

Part III – Sight

Rod and cone **photoreceptors** detect light stimuli and are located in the retina toward the back of the eyeball. They are more densely packed toward the center of the retina, where vision is the most acute. However, at the point where the optic nerve enters the eyeball, there are no receptors causing a blind spot in each eye.

1. Find your blind spot by holding this sheet of paper (or a 3 x 5 card given to you by your instructor) about a foot from your face, closing your left eye and looking at the "X" with your right eye. Slowly bring the paper toward you until the triangle disappears.

2. When closing your left eye, how far away (in mm or cm) from your right eye was the paper when the triangle disappeared? Record this in Table 4 in the results section.

3. Flip the paper or 3 × 5 card over so that the triangle is on the left and repeat the procedure for your left eye.

4. When closing your right eye, how far away (in millimeters) from your left eye was the paper when the triangle disappeared? Record this in Table 4 in the results section.

5. Have another group member repeat the experiment.

Part IV – Hearing

The sense of hearing is also mediated by **mechanoreceptors** like the touch sensation. However, the hearing mechanoreceptors are tiny hair cells found deep in your inner ear. Because of this similarity you can think of hearing as touch at a distance.

Hearing Sensitivity

1. Strike a tuning fork and hold it near one ear of your subject.
2. Slowly move the tuning fork away from the subject's ear until they can't hear the sound any more.
3. Measure how far from your subject's ear the sound can be heard. Record your measurements in Table 5.
4. Repeat with the other ear.
5. Change roles with your partner and repeat the experiment.

Localization of Hearing

1. Have the subject keep their eyes closed and plug one ear with a finger.
2. Strike a tuning fork and hold it above, below, in front, or behind their ear within the range they can hear the sound. Your goal is to test whether your subject can determine the location of the sound so keep them guessing!
3. Ask the subject whether the sound is above, below, in front of, or behind their ear.
4. If your subject guessed right put a "+" sign in Table 6 if she/he got it wrong put a "−" sign in Table 6.
5. Repeat this procedure two times for the first ear.
6. Switch ears and repeat steps 1 to 5.
7. Change roles with your partner and repeat the experiment.

Part V – Reaction Time

Unlike the other activities this activity is designed to measure your reaction time to something that you see. A reaction requires perception of the stimulus by the sensory receptor, followed by processing by the brain, and finally a response by your muscles (motor system).

1. Get a meter stick (same as a yard stick).
2. Hold the ruler near the end (highest number) and let it hang down.
3. Have your subject put his or her hand at the bottom of the ruler ready to grab the ruler (however, they should not be touching the ruler). See the figure to the right.

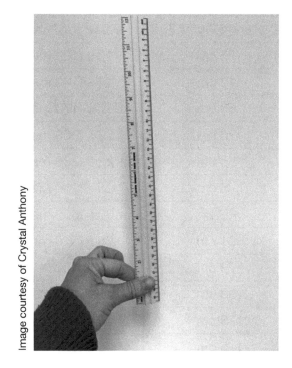
Image courtesy of Crystal Anthony

4. Tell your subject that you will drop the ruler sometime within the next 5 seconds and that they are supposed to catch the ruler as fast as they can after it is dropped.

5. Drop the ruler and record the level (in centimeters) at which they catch the ruler in Table 7 in the results section.

6. Test the same person 5 more times (vary the time of dropping the ruler within the 5 second "drop-zone" so the other person cannot guess when you will drop the ruler).

7. Change roles with your partner and repeat the experiment.

8. After your done take the average distance for all of your trials and convert your average distance to a reaction time using the table.

Here is a table to convert the distance on the ruler to reaction time. For example, if you caught the ruler at the 8 inch mark, then your reaction time is equal to 0.20 second (200 ms). Remember that there are 1000 ms (ms) in 1 second.

Distance Ruler Fell	Reaction Time
5 cm	10 s (100 ms)
10 cm	14 s (140 ms)
15 cm	17 s (170 ms)
20 cm	20 s (200 ms)
5 cm	23 s (230 ms)
5 cm	25 s (250 ms)
43 cm	30 s (300 ms)
61 cm	35 s (350 ms)
79 cm	40 s (400 ms)
99 cm	45 s (450 ms)
123 cm	50 s (500 ms)
175 cm	60 s (600 ms)

Answer Sheet

Lab 20: Senses

Name: _____   Section: _____

Part I – Taste

Tongue Map

U   U   U   U

Sour   Sweet   Salty   Bitter

1. Which taste was the easiest to map? Hardest?

TABLE 1 Salt Reaction Time

Salt Source	Time Interval to Taste (seconds)
Salt Crystals	
Salt Solution	

2. Were the two time intervals different? If so why?

TABLE 2 Localization of Touch in Three Locations

Partner	Back of Hand		Forearm		Palm	
	Trial #1 (mm)	Trial #2 (mm)	Trial #1 (mm)	Trial #2 (mm)	Trial #1 (mm)	Trial #2 (mm)
#1						
#2						

3. At which location was the localization of touch by the subject closest to the touch stimulus? At which location was the subject farthest off?

TABLE 3 Density of Mechanoreceptors in Three Locations

Back of hand						
	Partner #1			Partner #2		
Distance	Trial			Trial		
	1	2	3	1	2	3
20 mm						
15 mm						
10 mm						
5 mm						

Forearm						
	Partner #1			Partner #2		
Distance	Trial			Trial		
	1	2	3	1	2	3
20 mm						
15 mm						
10 mm						
5 mm						

Palm						
	Partner #1			Partner #2		
Distance	Trial			Trial		
	1	2	3	1	2	3
20 mm						
15 mm						
10 mm						
5 mm						

4. Based on your results above which area had the highest density of touch receptors (i.e., was the best at discriminate one point versus two points)? Which has the lowest density?

5. Do your results about the density of touch receptors in these different areas correspond with your results about the localization of touch in these areas? Explain.

6. Why do you think it would be advantageous (from an evolutionary perspective) to have more touch receptors in some areas of the body than in other areas?

TABLE 4 Distance at Which the Blind Spot is Observed

Partner #1		Partner #2	
Left Eye (cm)	Right Eye (cm)	Left Eye (cm)	Right Eye (cm)

7. Was the distance at which your blind spot was observed different for each eye? If so, what about your eyes do you think caused this difference? (Hint – it's the same thing that causes near sightedness and far sightedness, see your textbook for assistance.)

TABLE 5 The Distance Sound can be Heard at with Each Ear

Partner #1		Partner #2	
Left Ear (cm)	Right Ear (cm)	Left Ear (cm)	Right Ear (cm)

8. Was one ear more sensitive than the other? If so what do you think is different about your ears? Think about differences in the # of inner ear hair cells or possibly the shape or your ears.

TABLE 6 Localization of Sound by Both Ears

Distance	Partner #1			Partner #2		
	Trial			Trial		
	1	2	3	1	2	3
Left Ear						
Right Ear						

9. Was there a difference between your ability to localize the source of sound between your left ear and right ear? Did the difference correspond with differences seen for the sensitivity of hearing?

10. Based on your results above compare and contrast the sensitivity of hearing and touch. Both senses use mechanoreceptors but does one seem to be more sensitive than the other. Explain.

TABLE 7 Reaction Time for Responding to a Falling Ruler

Partner	Distance Ruler Fell (in cm)					Average	Reaction Time
	Trial #						
	1	2	3	4	5		
# 1							
# 2							

11. Did your reaction time improve during the experiment? If so why? Do you think that you could keep improving with more practice or is there a limit? Explain.

12. Describe step by step from your eye to the muscles in your hand what happened during the reaction.

13. Nerve impulses travel down an axon at ~1000 cm/ms which is much faster than your reaction time. Why is your reaction time slower than a nerve impulse? (Hint - Think about what has to happen at each step above.)

Laboratory 21

Homeostasis

Name: _____   Section: _____

Objectives

- Define homeostasis, intracellular fluid, extracellular fluid, plasma, interstitial fluid, homeostatic mechanisms, feedback systems
- Describe the role of sensors, integrating centers, and effectors in homeostatic feedback loops

Introduction

Your external environment is constantly changing. For example, air temperature change throughout the day and throughout the year. However, the body must maintain an internal environment that is very constant, not only in its temperature but in all of its other functional and structural characteristics. Maintenance of a stable internal environment is called **homeostasis**.

The internal environment is mostly fluid. Body fluid plays a very important role in homeostasis. Its major compartments are summarized in Table 21.1. Body fluid is divided into two major compartments, the intracellular fluid and the extracellular fluid. The extracellular fluid is further divided into interstitial fluid and blood plasma. There is ~44 L of fluid in the body.

TABLE 21.1 Fluid Compartments

Body Fluid	Definition	Function	Average Volume in the Body (L)
Intracellular	All of the fluid in cells	Contains dissolved substances needed for use by the cells.	29
Extracellular	All of the fluid outside the cell	Contains dissolved substances, contribute to regulation and communication of cells	15
• Interstitial	All of the fluid between the cells and the blood vessels	Diffusion of dissolved substances between cells and blood	12
• Plasma	All of the fluid portion of the blood	Transport of dissolved substances throughout body	3

Part I – Homeostatic Mechanisms and Feedback Systems (Loops)

Homeostatic mechanisms functions throughout the body to keep the chemistry of the body fluids constant and to maintain the various structures of the body. In order to understand Homeostasis, it is first important to understand the main functions of the various body systems.

Systems	Vital Functions
Integumentary	
Nervous	
Endocrine	
Skeletal	
Muscular	
Circulatory	
Lymphatic	
Respiratory	
Digestive	
Urinary	
Reproductive	

Feedback Systems

Regulation of Body Temperature is a classic example of a feedback system. All homeostatic mechanisms involve feedback loops (Figure 21.1). Negative Feedback loops are the most common, bringing things back to baseline. A **sensor (a receptor)** receives a stimulus – such as a rise or fall in body temperature. The sensor then sends the information about the change to the **integrating center (the brain)**. The integrating sends information to an **effector (organ, muscle, gland)**, which reverses the direction of the stimulus back to normal.

From your own experience and Figure 21.1, describe the effector that is activated when the body temperature gets too hot, what the response is and the organ system involved to cool the body down.

(a) Effector _____

(b) Response _____

(c) Name the involved organ system _____

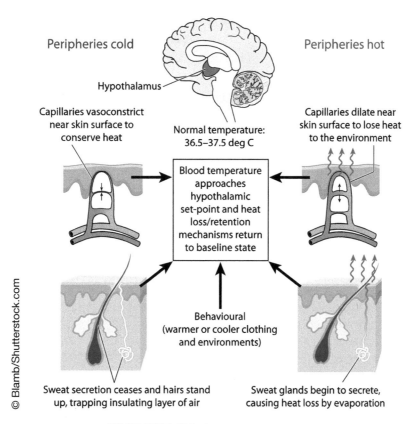

FIGURE 21.1 Thermoregulation.

Part II – The Effects of Exercise on Homeostasis

(A) Purpose

To discover the effect that various levels of exercise have on specific body parameters.

(B) Hypothesis

Write one hypothesis describing how you predict the five parameters below will change as the subject exercises.

- change in skin color on arms and face
- perspiration level
- external body temperature
- heart rate
- breathing rate

Hypothesis:

(C) Materials: Thermometer and stopwatch.

(D) Introduction

Exercise causes many homeostatic factors to kick in, in an effort to maintain internal homeostasis. How exercise affects some of these homeostatic factors can be determined by measuring and observing certain parameters such as:

- change in skin color on arms and face
- perspiration level
- external body temperature
- heart rate
- breathing rate

In the following lab 1 member of your group of will exercise for 8 minutes by skipping in place. The parameters listed above will be recorded at rest, 2, 4, 6, and 8 minutes, and 1 minute after exercise has stopped. The subject should stop just long enough for the needed measurements to be taken. ALL MEMBERS MUST HELP!!! Record all data in the table provided. The final lab report you turn in should follow the format that I have provided to you. In addition, your report is to include a graph for each of the 3 **measured** parameters (i.e., body temperature, heart rate, and breathing rate). The questions at the end of the lab should be a part of your result and discussion sections.

(E) Procedure

1. Each group should obtain: a thermometer and a stopwatch.
2. Record the RESTING observations and values of your subject for each of the five parameters.
    a. Record normal skin color of hands and face (i.e., pale, pink, red, etc.)
    b. Record normal perspiration level (i.e., none, mild, medium, high, etc.)
    c. Record external body temperature by placing the thermometer under the subjects arm pit for 1 minute (note: measurements should be taken directly from the skin).
    d. Determine the breathing rate by counting the number of breaths taken in 10 seconds and multiply by 6.
    e. Determine the heart rate by counting the number of beats in 10 seconds and multiply by 6. Use the radial artery at the wrist.
3. Have your subject begin to skip in place. Please note your subject should be sure to exercise at a level they can maintain for the entire 8 minutes.
4. Take your subjects parameter readings using the same techniques described above at the 2, 4, 6, and 8-minute time markers. Be sure to take final readings 1 minute after your subject has stopped exercising.
5. Record all of your parameter readings in the table provided.
6. After cleaning your thermometer, return it along with the other lab materials to the bin.

(F) Observations

	Body Color	Perspiration Level	Body Temp (Celsius) Wait 30 seconds for reading	Heart Rate (beats/min) Measure how many in 10 seconds and multiply by 6	Breathing Rate (breaths/min) Measure how many in 10 seconds and multiply by 6
REST					
2 min of exercise					
4 min of exercise					
6 min of exercise					
8 min of exercise					
1 min after exercise					

(G) Analysis

1. Plot your measured data on 3 graphs below. You should plot 1 graph each for body temperature, heart rate, and breathing rate versus the duration of exercise (i.e., 0 (rest), 2, 4, 6, 8, 9 minutes).

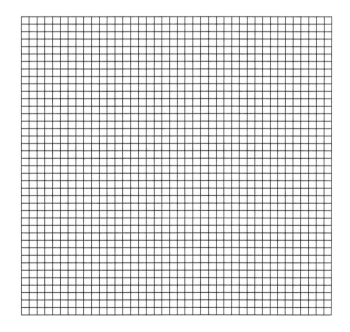

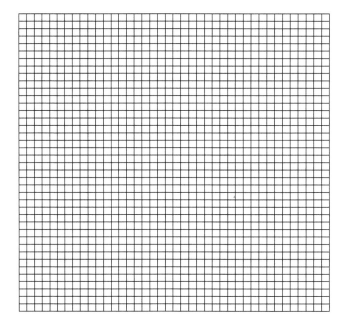

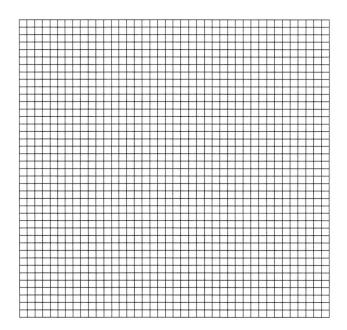

2. Write a paragraph describing your results for each of the 5 parameters in regards to homeostasis. Be specific, analyze, and reference your data.

(H) Discussion Questions: You may need to use your book to answer some of these questions

1. What are the changes you observed in body color and perspiration level in response to? How do these changes contribute to the maintenance of homeostasis?

2. Why do you think a change in body temperature occurs? What mechanisms does your body use to maintain its homeostatic temperature?

3. Why does an increase in heart rate and breathing rate accompany exercise?

4. By studying your parameter measurements after exercise has stopped, what conclusions can you make about your body's ability to maintain homeostasis?

Laboratory 22

Reproduction and Embryology

Objectives

- Define Fertilization, Cleavage, Gastrulation, Organogenesis
- Describe, identify and label the microscopic structures of the ovaries and testes
- Describe, identify and label the embryological stages of development

Introduction

Animal development has six stages (see diagram).

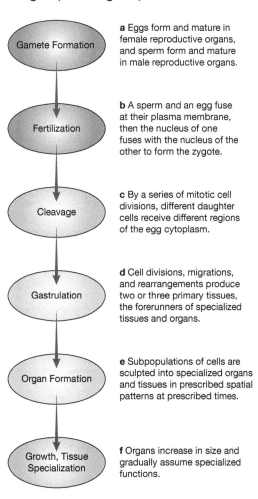

a Eggs form and mature in female reproductive organs, and sperm form and mature in male reproductive organs.

b A sperm and an egg fuse at their plasma membrane, then the nucleus of one fuses with the nucleus of the other to form the zygote.

c By a series of mitotic cell divisions, different daughter cells receive different regions of the egg cytoplasm.

d Cell divisions, migrations, and rearrangements produce two or three primary tissues, the forerunners of specialized tissues and organs.

e Subpopulations of cells are sculpted into specialized organs and tissues in prescribed spatial patterns at prescribed times.

f Organs increase in size and gradually assume specialized functions.

FIGURE 22.1 Generalized scheme and control of animal development.

In this lab, you will start by viewing and drawing the male and female gametes (using slides of the *ovaries* and *testes*).

You will then view and draw the stages of embryological development.

A **zygote** is formed after fertilization occurs (union of the male and female gametes).

Cleavage is a special type of cell division that occurs first in the zygote and then in the cells formed by successive cleavages, the blastomeres. Unlike typical cell division, there is no intervening period of cytoplasmic growth between mitotic divisions. Thus, the blastomeres become smaller and smaller. After a number of cleavages, the blastomeres form a solid ball of cells called the *morula*. The formation of a hollow ball of cells, called the *blastula* marks the end of cleavage. (Because there is no cytoplasmic growth, the size of the blastula is only slightly larger than that of the zygote.)

Gastrulation cells move and migrate (morphogenesis). Primary Germ layers (tissue layers) develop; called ectoderm, mesoderm and endoderm. All organs can be traced back to one pf the primary germ layers.

Organogenesis is the time of organ formation.

During **growth and tissue specialization**, the sixth stage, organs grow in size and acquire the specialized functions necessary for an independent life.

Lab 22: Pre-Lab Quiz

Name: _____

Read through the entire lab and answer these questions before coming to class.

1. The ovaries contain _____.
   a. Gametes
   b. Zygotes
   c. Blastulas
   d. Spermatocytes

2. The cell that is formed after fertilization occurs is called the _____.
   a. Blastula
   b. Gastrula
   c. Morula
   d. Zygote

3. A solid ball of cells is called the _____.
   a. Blastula
   b. Gastrula
   c. Morula
   d. Zygote

4. A hollow ball of cells is called the _____.
   a. Blastula
   b. Gastrula
   c. Morula
   d. Zygote

5. During specialization _____ cells.
   a. Have specific functions
   b. Have multiple functions
   c. Are in the S phase
   d. Undergo apoptosis

6. Organogenesis refers to:
   a. Fetal development
   b. Infant development
   c. Organ development
   d. Arrested development

7. In this lab you will need to label 2 items.
   a. For every drawing
   b. For some of the drawings
   c. For none of the drawings
   d. If you feel like it

Answer Sheet

Lab 22: Reproduction and Embryology

Name: _____   Section: _____

1. Male and Female Reproductive organs containing gametes

Women ovaries have follicles that contain the Oocytes. Due to hormonal signaling, Oogenesis occurs and one follicle outgrows the other. A Secondary oocyte (egg cell) is released during ovulation.

Male testes have seminiferous tubules that contain sperm cells that have been released through spermatogenesis, this can be seen in the tissue of the testis.

With your compound microscope, observe a prepared slide of the female ovary and the male testis. **Draw and label two structures**.

Ovary

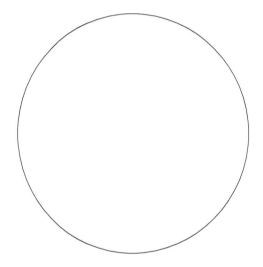

Magnification_____

Testis

Magnification_____

2. Cleavage In the Sea Star

With your compound microscope, observe a prepared slide with whole mounts of early sea star embryos. **Find the Zygote, 2-cell stage, 4-cell stage, 8-cell stage, morula, and blastula, and draw them**. Be sure to adjust the fine-focus knob to see the three dimensional aspects of these stages. The blastula is a hollow ball of flagellated blastomeres surrounding a cavity called the **blastocoel. Label the blastomeres and blastocoel in your drawing of the sea star blastula**.

Zygote

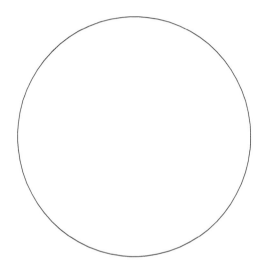

Magnification_____

2 Cell

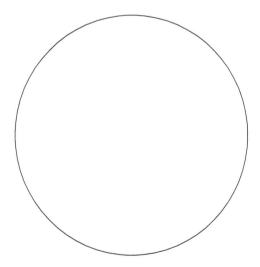

Magnification_____

4 Cell

Magnification _____

Morula

Magnification _____

Blastula

Magnification_____

Laboratory 23

Cell Division

Objectives

- Define the following; chromosome, centromere, sister chromatid, chromatin, cytokinesis, prophase, metaphase, anaphase, telophase, and interphase
- Describe why nuclear and cytoplasmic division are important
- Locate and identify the principal stages of mitosis
- Diagram and label the principle stages of mitosis
- Describe how cytokinesis differs between plant and animal cells

Introduction

The **cell theory** states that all cells come from pre-existing cells. Eukaryotic organisms reproduce their cells through mitosis and meiosis. The process of **mitosis** divides one parent cell into two daughter cells. Each daughter cell has two copies of every chromosome (2N) just like the original parent cell. In humans, mitosis happens throughout the body to maintain our organs and tissues. In plants mitosis is mostly restricted to the **apical meristems** at the tips of stems and roots. The process of **meiosis** only happens in reproductive organs (ovaries and testes in humans; ovaries and anthers in flowering plants) where one parent cell divides into four gamete cells. Gametes are the reproductive cells (eggs and sperm in humans; eggs and pollen in plants) that are used for mating. Each gamete only has one copy of each chromosome (1N).

DNA is the genetic material of cells that is used to make proteins. In resting cell's DNA is stored in the nucleus as long strands called **chromatin**. All cells need DNA to function so during the **S phase of interphase** prior to mitosis and meiosis the DNA in the parent cell is copied. At the beginning of mitosis and meiosis the chromatin strands condense into compact **chromosomes** that are visible through a microscope as structures that look like X's in the center of the cell. Each chromosome is made up of two copies or **sister chromatids** held together by a **centromere** (each side of the X is one copy and the centromere is the constriction point). At the end of mitosis and meiosis the process of **cytokinesis** splits the parent cell into two daughter cells. In animal cells the plasma membrane is pinched apart by the cleavage furrow that forms in the center of the cell. In plants cells vesicles line up at the center of the cell and form a **cell plate** that becomes a new cell wall between the cells.

Use your text, laboratory models, and charts to review the stages of mitosis and meiosis. Don't rush looking at the slides today. It takes time and effort to find and identify each step in the mitotic cycle. Remember, the naming and description of each step is artificial as mitosis is actually a continuous process. The numerous steps in texts are there to help.

Prior to making your microscopic observations make sure that you have an understanding of what is happening in each stage of mitosis. Remember, while what you observe on a slide will look similar to the drawing or photograph in a text, the stages are artificial and there will be slight differences between a textbook presentation and what you actually observe.

The cell cycle in plant cells (Onion Roots)

Nuclear and cell divisions in plants are, for the most part, localized in specialized regions called meristems. Meristems are regions of active growth.

Use Google Images search "Onion Root tip mitosis" use the images you find to familiarize yourself with the stages of mitosis and so you can recognize them.

Draw and label cells in the onion root tip at interphase and each stage of mitosis (Prophase, Metaphase, Anaphase, Telophase) in the labeled circles on the slide drawing sheets at the end of the lab. **Make sure to label the chromosomes, cell wall, and nucleus if present.** Figure 23.1 shows examples of onion cells at the different stages you will see at 400× total magnification.

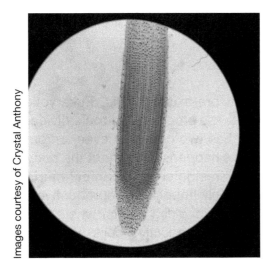

Images courtesy of Crystal Anthony

FIGURE 23.1 Examples of onion cells at the different stages you will see at 400× total magnification.

Lab 23: Pre-Lab Questions

Name: _____

Read through **all** of the lab handout and answers these questions **before** coming to class.

1. _____ theory states that cells come from pre-existing cells.
   a. Meiosis
   b. Cell
   c. Mitosis
   d. Chromosome

2. Which of these divides one parent cell into two daughter cells?
   a. Meiosis
   b. Fertilization
   c. Mitosis
   d. Cleavage

3. Which of these divides one parent cell into four gamete cells?
   a. Meiosis
   b. Fertilization
   c. Mitosis
   d. Cleavage

4. Which of these is not a stage of mitosis?
   a. Prephase
   b. Prophase
   c. Anaphase
   d. Telophase

5. Which of these stages happens before mitosis?
   a. Prophase
   b. Metaphase
   c. Interphase
   d. Telophase

6. In resting cells, DNA is stored in the nucleus as long strands called _____.
   a. Centromeres
   b. Sister Chromatids
   c. Nucleosomes
   d. Chromatin

7. Which of these is the earliest stage of mitosis?
   a. Anaphase
   b. Telophase
   c. Cytokinesis
   d. Prophase

8. Which of these is the stage when one cell is divided into two cells?
   a. Anaphase
   b. Telophase
   c. Cytokinesis
   d. Prophase

9. Which of these is the area of the onion root where mitosis happens?
   a. Root cap
   b. Apical meristem
   c. Blastula
   d. Cleavage

10. _____ is the special type of cell division that the zygote undergoes to produce a blastula stage embryo.
    a. Root cap
    b. Apical meristem
    c. Blastula
    d. Cleavage

Answer Sheet

Lab 23: Cell Division

Name: _____ Section: _____

Part I – The cell cycle and mitosis

a. Describe what is happening in each stage of mitosis. Refer to your textbook for help.

Interphase _____
_____
_____
_____

Prophase _____
_____
_____
_____

Metaphase _____
_____
_____
_____

Anaphase _____
_____
_____
_____

Telophase _____
_____
_____
_____

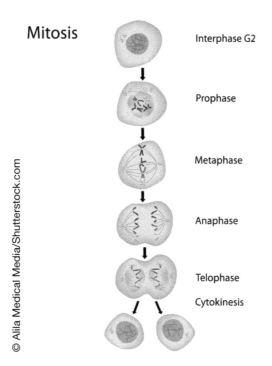

b. Microscope Drawing Sheets

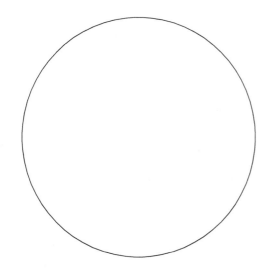

Drawing of an onion cell undergoing
Interphase
Label the **chromosomes, cell wall, and nucleus**

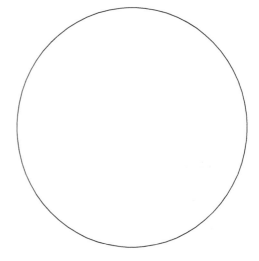

Drawing of an onion cell undergoing
Prophase
Label the **chromosomes** and **cell wall**

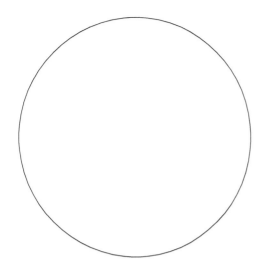

Drawing of an onion cell undergoing
Metaphase
Label the **chromosomes** and **cell wall**

Drawing of an onion cell undergoing
Anaphase
Label the **chromosomes, cell wall**.

Part II – Meiosis

a. Briefly describe what is happening in each stage of meiosis. Refer to your textbook for help.

Interphase I _____

_____

Prophase I _____

_____

Metaphase I _____

Anaphase I _____

_____

Telophase I _____

_____

Prophase II _____

_____

Metaphase II _____

Anaphase II _____

_____

Telophase II _____
_____

b. Meiosis occurs to create the haploid gametes cells (egg and sperm) that are required for sexual reproduction. As you probably know gametes are only produced in the gonads (ovaries and testes) of humans and other higher animals.

What about plants? Do plants have sex? Indeed they do. The gametes of flowering plants are produced in the flower itself. The sperm are produced and packaged into pollen in the male part of the flower called the anther and the egg is produced in the female part of the flower called the ovule which is located inside of the ovary that becomes the fruit. In this part of the lab you will be observing stages of meiosis happening in the anther of the lily flower.

Meiosis proceeds through stages that are similar to mitosis but meiosis repeats the process twice (meiosis I and meiosis II) to make four cells instead of just one cell. Please refer to your textbook for a more complete description of the stages of meiosis.

Table 1 Observe the Demonstrations Slides Numbered 1 through 4 and **Name the Stages of Meiosis Below**. Each Stage Corresponds to One and Only One of the Demonstrations Slides

Stage of Meiosis	Demonstration Slide #
	1
	2
	3
	4

Part III – Human Chromosomes

View the Demonstration slides of human chromosomes. These cells are stained with a special dye that doesn't stain the cell membrane or nucleus but targets just the chromosomes.

Questions

1. Compare and contrast mitosis and meiosis.

2. What type of issues might occur in the integumentary system if there were problems with mitosis?

3. Where does mitosis happen in humans? Where does meiosis happen? Use your textbook for assistance. Be specific.

4. What happens to the chromosomes in the S phase of the cell cycle? If a cell skipped S phase what would happen during mitosis?

Laboratory 24

"Genes in a Bottle" Cheek Cell DNA Extraction – Capture Your Genetic Essence in a Bottle

Objectives

- Describe the structure of DNA and where it is found
- Describe the procedure for extracting DNA
- Extract your DNA from your cheek cells

Introduction

What is DNA and what does it do?
Deoxyribonucleic acid (DNA) is a molecule present in all living things, including bacteria, plants, and animals. DNA carries genetic information that is inherited, or passed down from parents to offspring. It is responsible for determining a person's hair, eye, and skin color, facial features, complexion, height, blood type, and just about everything else that makes an individual unique. But it also contains all the information about your body that is the same in all human beings. In other words, your DNA is like a blueprint for your entire physical growth and development. Your DNA blueprint is a combination of half of your mother's and half of your father's DNA, which is why you have some features from each of your parents.

DNA contains four chemical units, referred to by the first letters in their names: **A** (adenine), **G** (guanine), **T** (thymine), and **C** (cytosine). These four DNA "letters" make up a code for genetic information. The letters of the DNA code are similar to the letters of our alphabet. The 26 letters in our English alphabet spell words, which can be arranged in infinite ways to create messages and information. Similarly, the 4 chemical letters of DNA are organized to make messages, called **genes**, that can be understood by cells. These genes contain the information to make **proteins**, which are responsible for almost all of your body's structures and functions. A gene is like a recipe, since it contains the all the information needed to make a protein.

Your DNA sequence is the particular arrangement or order of the chemical letters within your complete DNA collection, or **genome**. Scientists have determined that human DNA sequences are 99.9% identical. It is the <0.1% sequence variation from person to person that makes each of us unique. In other words, what makes you different from your classmate is an occasional difference in the letters of your genomes.

Where is DNA found?

The basic units of an organism's body are cells—they make up all of your tissues and organs (e.g., muscles, brain, digestive system, skin, glands, etc.) Cells are compartments with membranes, made of protein and lipids (fats) that keep them separate from other cells. Within cells are further compartments with specialized functions. One compartment, called the **nucleus**, is like the cell's control headquarters and contains the DNA molecules, which are the master instructions for the functions of the cell. The DNA is organized into 46 tightly coiled structures called chromosomes. Every time a cell divides to make two identical new cells—for growth, repair, or reproduction—the chromosomes are copied, ensuring that the new cells will receive a full copy of the genetic blueprint for the organism.

What does DNA look like?

At the molecular level, DNA looks like a twisted ladder or a spiral staircase. The ladder actually contains two strands of DNA, with pairs of the chemical letters **A**, **G**, **T**, and **C** forming the rungs. This structure is called a DNA **double helix** because of the spiral, or helical form made by the two DNA strands. Each strand of DNA is very long and thin and is coiled very tightly to make it fit into the cell's nucleus. If all 46 human chromosomes from a cell were uncoiled and placed end to end, they would make a string of DNA that is 2 meters long and only 2 nanometers (2 billionths of a meter) wide!

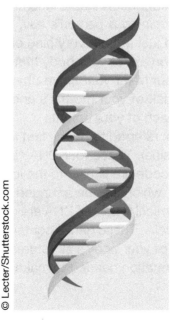

FIGURE 24.1 An illustration of DNA (Deoxyribonucleic Acid).

How can we make DNA visible?

Collect cells

To see your DNA, you will collect cells, break them open, and condense the DNA from all of the cells together. You can collect thousands of cells from the inside of your mouth just by scraping it gently and thoroughly with a brush. The type of cells that line your mouth divides very often, coming off easily as new cells replace them continuously. In fact, these cells are coming off and being replaced every time you chew and eat food.

Break open (lyse) the cells

Once you have collected your cells, the cells need to be broken open to release the DNA. Detergent will dissolve the membranes of your cells, just like dishwashing detergent dissolves fats and proteins from a greasy pan, because cell and nuclear membranes are composed of fats and proteins. Dissolving the membranes results in the release of the DNA. The process of breaking open the cells is called lysis, and the solution containing the detergent is called **lysis buffer**.

Remove proteins

DNA is packaged tightly around proteins. Like spools for thread, these proteins keep the DNA tightly wound and organized so that it doesn't get tangled inside the nucleus. For you to see the DNA, it helps to remove the proteins so that the DNA can first loosen and expand, then collect into a mass with the DNA from all the other cells. You will incubate your lysed cheek cells with **protease**, which breaks down proteins so that they can no longer bind DNA. Protease is an **enzyme**, or protein machine, that works best at 50°C, which is the temperature of slightly hot water. The protease chews up the proteins associated with the DNA and also helps digest any remaining cell or nuclear membrane proteins.

Condense the DNA

Strands of DNA are so thin that it is not possible to see them when they are dissolved in solution. Think of the long, thin strands of DNA as fine white thread. If one long piece of thread were stretched across the room, it would be difficult to see. To make the thread more visible, you could collect it all together and pile it on the floor. In this laboratory experiment, you will use salt and cold alcohol to bring the DNA out of solution, or **precipitate** it. Salt and cold alcohol create a condition in which DNA doesn't stay in solution, so the DNA clumps together and becomes a solid mass that you can see.

What does precipitated DNA look like?

Like salt or sugar, DNA is colorless when it is dissolved in liquid, but is white when it precipitates in enough quantity to see. As it precipitates, it appears as very fine white strands suspended in liquid. The strands are somewhat fragile—like very thin noodles, they can break if handled roughly. Also, if a mass of precipitated DNA is pulled out of its surrounding liquid, it will clump together, much like cooked noodles will clump together when they are pulled out of their liquid.

After the Exploration – Expected Results

A slimy white material will precipitate at the interface of the ethanol and filtrate layers. This material consists of clumped-together DNA strands and some protein.

Lab 24: Pre-Lab Questions

Name: _____

Read through **all** of the lab handout and answers these questions **before** coming to class to prepare for the Pre-lab quiz.

1. The structure of DNA is a double _____.
   a. Loop
   b. Circle
   c. Helix
   d. Spiral

2. You will be extracting DNA from your _____ cells.
   a. Blood
   b. Muscle
   c. Cheek
   d. Hair

3. Each of your body cells has _____ chromosomes.
   a. 2
   b. 10
   c. 24
   d. 46

4. Which of the following breaks open the cells to release the DNA?
   a. Lysis buffer
   b. Protease enzymes
   c. Ethanol alcohol
   d. Pipetting

5. Which of the following degrades the proteins associated with DNA?
   a. Lysis buffer
   b. Protease enzymes
   c. Ethanol alcohol
   d. Pipetting

6. DNA becomes visible after it is _____.
   a. Lysed
   b. Proteased
   c. Precipitated
   d. Pipetted

7. Which of the following precipitates the DNA?
   a. Lysis buffer
   b. Protease enzymes
   c. Ethanol alcohol
   d. Pipetting

8. How many milliliters of Gatorade will you swish in your mouth to collect the cheek cells?
   a. 1 mL
   b. 3 mL
   c. 5 mL
   d. 15 mL

9. The complete collection of DNA and genes is called the _____.
   a. Proteome
   b. DNAome
   c. Genome
   d. Chromosome

10. In which organelles is DNA found (there is more than one correct answer!)?
    a. Mitochondria
    b. Endoplasmic reticulum
    c. Nucleus
    d. Golgi apparatus

Procedures

Quick Guide for DNA Extraction and Precipitation

1. Obtain a small cup containing 3 mL of Gatorade from your instructor. Label the tube with your initials
2. Gently chew the insides of your cheeks for 30 seconds. It is NOT helpful to draw blood!
3. Take the Gatorade from the cup into your mouth, and swish the Gatorade around vigorously for 30 seconds.
4. Expel the Gatorade into a paper/plastic cup
5. Carefully pour the solution from the cup into the 15 mL test tube
6. Obtain the tube of lysis buffer from your workstation, and add 2 mL of lysis buffer to your tube. There is a black mark labeling where 2 mL is located on the pipette.
7. Place the cap on the tube, and gently invert the tube 5 times (don't shake your tube!). Observe your tube – do you notice any changes? If you do, write them down.

8. Place your tube in a test tube rack or beaker in the water bath and incubate at 50°C for 10 minutes. Remove your tubes from the water bath.

Water bath

50°C for 10 min

9. Obtain the tube of cold alcohol from your instructor or at the common workstation. Holding your tube at a 45° angle, fill your tube with cold alcohol (your tube should almost be filled to the top with liquid). It will take repeated additions to fill your tube with cold alcohol using the disposable plastic transfer pipet.

10. Place your cap on your tube, and let it sit undisturbed for 5 minutes. Write down anything you observe happening in the tube.

11. With a disposable plastic transfer pipette, carefully transfer the precipitated DNA along with the alcohol solution into a small glass vial provided in the DNA necklace kit.

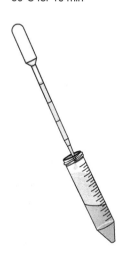

12. Firmly push the plastic stopper cap into the neck of the vial to seal the glass vial.

13. After the glue has dried, slip the waxed cord through the silver cap and tie the cord

14. Apply a small drop of glue into the inside of the silver cap. Apply a small amount of glue around the rim of the glass vial/plastic stopper cap. Do not apply too much glue as it may interfere with the drying process. A few of the newer necklaces do not require glue. Check with your instructor

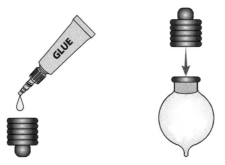

15. Place the silver cap onto the top of the glass vial and press down firmly for 30 seconds. Allow the glue to dry for 10 to 15 minutes and then check for a complete seal.

Congratulations, you've created your very own DNA necklace!

Clean-Up

Pour the remaining DNA/ethanol solution into the designated disposal can.
Rinse out your test tube and place it in the dirty dish container.
Return all of your lab supplies to the lab cart.
Wipe down your lab bench.

Lab 24: Genes in a Bottle End of Exercise Questions

Name: _____   Section: _____

1. Where is the DNA inside a eukaryotic cell located? Is there actually more than one location? Be specific.

2. Were you surprised by how much DNA came out of your cheek cells? How does all of that genetic material fit into your tiny cells? Use your textbook for assistance.

3. For each step in the DNA extraction process below indicate why the solutions that were added.

    Solution added – Gatorade (Protease)

    Reason step was performed –

    Solution added – Lysis Buffer

    Reason step was performed –

    Solution added – Cold Alcohol

    Reason step was performed –

Laboratory 25

Mendelian Inheritance

Objectives

- Define allele, genotype, gametes, heterozygous, homozygous, independent assortment, dominant, recessive, incomplete dominance, and sex-linked
- Solve genetic problems

Introduction

In 1866 an Austrian monk, Gregor Mendel, presented the results of painstaking experiments on the inheritance of the garden pea. Mendel discovered that his pea plants are **diploid** which means that they contain two copies of each gene for every trait. Each gene copy is called an **allele**. Mutation can change the sequence of DNA in a gene creating new alleles with different functions. For instance Mendel showed that there are two alleles for the flower color gene in pea plants, one allele makes the flower petals purple and the other allele make the flower petals white. Mendel went on to

show that when he crossed purple pea plants with white pea plants to make a pea plant with one copy of each allele the flower petals were still purple demonstrating that the purple allele is dominant over the recessive white allele. **Dominant** alleles like purple are given an upper case letter (P) whereas **recessive** alleles like white are given a lower case letter (p). The alleles an individual has for a gene are called its **genotype**. In diploid organisms the genotype is made up of two alleles and can be homozygous meaning that both the alleles are the same or **heterozygous** meaning that each allele is different. In our example pea plants that are homozygous for the dominant purple allele (PP) make purple flowers, pea plants that are heterozygous (Pp) also make purple flowers because the purple allele is dominant, and pea plants that are homozygous for the white allele (pp) make flowers with white petals.

Mendel also showed that the inheritance of the pea plant traits he was studying follows two laws. The **law of segregation** states that the allele pairs in the parents separate from each other during meiosis and produce a sperm or egg that carries only one allele for each trait. The **law of independent assortment** states that alleles for different genes sort out independently during meiosis creating all possible allele combinations. Together these laws showed that genetic inheritance followed the rules of probability making it possible to predict the likelihood of an offspring inheriting a trait from its parents. In this lab we will use a **punnet square** to help us calculate the probabilities of various outcomes for several different types of genetic crosses.

At the time of Mendel's discoveries his results were heard, but probably not understood. Finally around 1900, Mendel's work was rediscovered by geneticists that still couldn't figure out inheritance and his profound insights were finally appreciated. Today Mendel's laws seem elementary to modern-day geneticists, but its importance cannot be overstated. The principles generated by his pioneering experimentation are the foundation for genetic counseling so important today to families with health disorders having a genetic basis. It is also the framework for the modern research that is making inroads in treating diseases previously believed incurable.

Recent advances in molecular genetics have resulted in the production of insulin and human growth hormone by genetic engineering techniques. However, this new technology has not been without controversy. Genetically engineered diseased resistant crops, and even crops capable of withstanding temperatures that normally cause freezing, have met strong public opposition.

In the future you may be called upon to help make decisions about issues like these. To make an educated judgment, you must understand the basics, just as Mendel did. The genetics problems in this exercise should start you well on your way.

Inheritance Problems

Solving a heredity problem requires five basic steps.

1. Assign a symbol for each allele.
   Usually a capital letter is used for a dominant allele and a small letter is used for a recessive allele.

   Example: E = free earlobes which is dominant
   e = attached earlobes which is recessive

2. Determine the genotype of each parent and indicate a mating.
   If both parents are heterozygous, then the male is Ee, and the female is also Ee.
   Ee × Ee

3. Determine all the possible kinds of gametes each parent can produce.
   Remember that all gametes are haploid; therefore, they can have only one allele instead of the two present in the diploid cell. The male has two possible gametes: E and e. The female is the same.
   Male possible gametes: E and e
   Female possible gametes: E and e

4. Determine all the allele combinations that can result from the combining of gametes.
   Here we can use a Punnett square to determine the possibilities:

	E	e
E	EE	Ee
e	Ee	ee

5. Determine the genotype of each possible allele combination shown in the offspring.
   Phenotypes: three have free earlobes (EE, Ee); one has attached earlobes (ee).

   Genotype probabilities: ¼ is homozygous dominant, ½ is heterozygous dominant, and ¼ is homozygous recessive.

Lab 25: Pre-Lab Questions

Name: _____

Read through the lab handout and answers these questions **before** coming to class.

1. _____ is the study of how traits are inherited.
   a. Chemistry
   b. Ecology
   c. Genetics
   d. Phrenology

2. Meiosis produces haploid _____.
   a. Individuals
   b. Body cells
   c. Gametes
   d. Chromosomes

3. Mendel's law of _____ states that alleles for two different genes separate during meiosis and segregate independently.
   a. Independent segregation
   b. Independent assortment
   c. Independent mitosis
   d. Independent meiosis

4. Which of these is heterozygous?
   a. AA
   b. aa
   c. Aa

5. Which of these is homozygous for the recessive allele?
   a. AA
   b. aa
   c. Aa

6. What is the name of the table (or square) used to calculate the probabilities of offspring genotypes?
   a. Mendel square
   b. Genetic square
   c. Punnet square
   d. Darwin square

7. Which type of inheritance is affected by gender?
   a. Sex-linkage
   b. Incomplete dominance
   c. Complete Dominance
   d. Co-dominance (Multiple alleles)

8. In which type of inheritance is an intermediate phenotype observed?
   a. Sex-linkage
   b. Incomplete dominance
   c. Complete Dominance
   d. Co-dominance (Multiple alleles)

9. The observed trait an individual has is called the _____.
   a. Allele
   b. Gene
   c. Phenotype
   d. Genotype

10. Which type of inheritance has two or more completely dominant alleles?
    a. Sex-linkage
    b. Incomplete dominance
    c. Complete Dominance
    d. Co-dominance (Multiple alleles)

Answer Sheet

Lab 25: Mendelian Inheritance

Name: _____   Section: _____

Inheritance Problems: Single Factor

1. In humans, six fingers (F) is the dominant trait; five fingers (f) is the recessive trait. Assume both parents are heterozygous for six fingers.

   a. What is the **phenotype** of the father?

   b. What is the **phenotype** of the mother?

   c. What are the **genotypes** of the parents? Set up the cross and use a punnet square to determine the genotypes and phenotypes of the children.

   d. What is the probability of a 6 fingered child?

   e. What is the probability of a 5 fingered child?

2. If the **father is heterozygous for six fingers** and the **mother has five fingers**, what is the probability of their offspring having five fingers or six fingers? Show your work.

3. For this question use **W** to represent the white allele and **R** to represent the red allele. In certain flowers, color is inherited by alleles that show *incomplete dominance*. In such flowers, a cross between a homozygous red flower (RR) and a homozygous white flower (WW) always results in a heterozygous pink flower (WR). Neither the red allele nor the white allele dominates but instead an intermediate phenotype is seen in the heterozygote. **A cross is made between two pink flowers**. What is the probability of each of the colors (red, pink, and white) appearing in the offspring?

4. Use the information given in the previous problem. **A cross is made between a red flower and a pink flower**. What is the expected probability for the various colors?

Inheritance Problems: double factor

A double-factor cross is a genetic study in which two pairs of alleles are followed from the parental generation to the offspring. These problems are worked the same as single factor crosses, only you work with two different characteristics from each parent.

It is necessary to recognize that *independent assortment* occurs when two or more sets of alleles are involved. *Mendel's law of independent assortment* states that members of one allelic pair separate from each other independently of the members of other pairs of alleles. This happens during meiosis when the chromosomes that carry the alleles segregate. Mendel's Law of Independent Assortment applies only if the two pairs of alleles are located on different pairs of homologous chromosomes. This is an assumption we will use in double-factor crosses.

In the following problem, go through the basic steps of solving a double-factor genetic problem. Show each step of the process.

5. In humans, the allele for free earlobes dominates the allele for attached earlobes. The allele for dark hair dominates the allele for light hair. If **both parents are heterozygous for earlobe shape and hair color**, what genotypes and phenotypes can their offspring have and what is the probability of each genotype and phenotype?

   a. Assign a symbol for each allele.

   b. Determine the genotype of each parent and indicate a mating.

   c. Determine all the possible kinds of gametes each parent can produce.

   d. Determine all the allele combinations that can result from the combining of gametes. (draw and use a Punnett square)

   e. Determine the genotype and phenotype of each possible allele combination shown in the offspring.

	Phenotypes			
	(1)	(2)	(3)	(4)
Genotypes:				

Check your work with another group or your instructor before continuing with the rest of the problems.

6. In humans, a type of blindness is due to a dominant allele and normal vision is the result of a recessive allele. Migraine headaches are due to a dominant allele, and normal (no headaches) is recessive. A **male who is heterozygous for blindness and does not suffer from headaches** marries **a woman who has normal vision and is heterozygous for migraines**. Could they produce a child with normal vision who does not suffer from headaches? If yes, can the probability of such a child be determined? Show your work.

Sex-linked problem.

7. In humans, the condition for normal vision dominates color blindness; both alleles are linked to the X chromosome. **A normal male marries a color-blind female**. If they have a daughter, what is the chance she will have normal vision? What about a son? Show your work.

Co-dominance (Multiple alleles) problem.

8. In humans, there are three alleles for blood type: A, B, and O. The allele for blood type A and allele for blood type B show co-dominance. A person with both alleles has blood type AB. Both A and B dominate type O. **A man with alleles for blood types A and O marries a woman with alleles for blood type B and O**. List the types of blood their offspring could have and the probability of each blood type.

The following genetics problem requires some detective work to determine the genotypes.

9. Normal pigmentation (A) dominates no pigmentation (albino 5 aa). Dark hair coloring (D) dominates light hair coloring (d). **Two parents both with normal pigmentation, one with dark hair, and the other with light hair produce a child with dark hair and normal pigmentation (Child A), a child with light hair and normal pigmentation (Child B), and an albino child (Child C)**. Albinism is epistatic to hair color meaning that you can't tell the real hair color because the albino genotype makes everyone's hair white.

What are the genotypes for the parents? Determine the possible genotypes of the children first and use that information to figure out the parents genotypes. You cannot be sure of the children's genotype for certain traits so write out all of the possibilities. Child A has been done for you as an example.

Child	Pigmentation/genotype	Hair color/genotype
Child A	Normal Aa or AA	Dark Dd
Child B		
Child C		
Light-haired Parent		
Dark-haired Parent		

Practical Exam Sheets

Title: _____ Lab Practical # _____   **Name:** _____

1 _____
2 _____
3 _____
4 _____
5 _____
6 _____
7 _____
8 _____
9 _____
10 _____
11 _____
12 _____
13 _____
14 _____
15 _____
16 _____
17 _____
18 _____
19 _____
20 _____

Title: _____ Lab Practical # _____　　**Name:** _____

1. _____
2. _____
3. _____
4. _____
5. _____
6. _____
7. _____
8. _____
9. _____
10. _____
11. _____
12. _____
13. _____
14. _____
15. _____
16. _____
17. _____
18. _____
19. _____
20. _____

Title: _____Lab Practical #_____ **Name:** _____

1. _____
2. _____
3. _____
4. _____
5. _____
6. _____
7. _____
8. _____
9. _____
10. _____
11. _____
12. _____
13. _____
14. _____
15. _____
16. _____
17. _____
18. _____
19. _____
20. _____